Tooling for Composite Aerospace Structures

Tooling for Composite Aerospace Structures

Manufacturing and Applications

Zeaid Hasan
Boom Supersonic, Denver, Colorado, USA

ELSEVIER
Butterworth-Heinemann
An imprint of Elsevier

Butterworth-Heinemann is an imprint of Elsevier
The Boulevard, Langford Lane, Kidlington, Oxford OX5 1GB, United Kingdom
50 Hampshire Street, 5th Floor, Cambridge, MA 02139, United States

Library of Congress Cataloging-in-Publication Data
A catalog record for this book is available from the Library of Congress

British Library Cataloguing-in-Publication Data
A catalogue record for this book is available from the British Library

ISBN: 978-0-12-819957-2

For information on all Butterworth-Heinemann publications
visit our website at https://www.elsevier.com/books-and-journals

Publisher: Matthew Deans
Acquisitions Editor: Carrie Bolger
Editorial Project Manager: Joshua Mearns
Production Project Manager: Nirmala Arumugam
Designer: Christian J. Bilbow

Typeset by Thomson Digital

To my mother Yusra Shalen and father Fouad Hasan
And the people of Gaza

Contents

Disclaimer

The information provided in this book is designed to provide helpful information to the subjects discussed. This book is not meant to be used as a process specification for any specific product, rather should be used as a general guideline to inform engineers during the design, analysis, and fabrication phases. The author and publisher are not responsible for any damages or negative consequences to any person reading or following the information in this book.

Preface

This book addresses the use of composite materials manufacturing focusing on the tooling aspects used during lamination and assembly. It is intended for use as a primary textbook for a course in composite materials manufacturing and analysis or as a supplementary text in manufacturing engineering courses, and as a resource for engineers in the field of aerospace manufacturing, design, materials, and stress analysis.

The main characteristics of this book are:

- It covers the essential aspects related to manufacturing of composite parts with an emphasis on tools. Since tooling is considered one of the primary drivers of cost and schedule in every program, it is worthwhile spending the time early on to ensure that all critical details are covered.
- It assumes that the reader's goal is to understand the entire life cycle of tooling starting from the material down to how it will be used on the production floor. The book is presented in a way as to provide engineers with a methodical way to think about the tooling details. It also offers a way to achieve a suitable balance of cost, schedule, and quality. It is not oriented toward details of design for manufacturing although this book should be used as a supplement to evaluate the critical details associated with tooling.
- It presents a wide range of materials used for tooling and discusses manufacturing methods used for composite parts and the tooling aspects associated with each. This will allow students and engineers to understand the critical details and the different manufacturing methods that exist and use depending on the application for optimal performance. This will also provide engineer's with guidance on what type of materials and manufacturing methods to use depending on the cost, schedule, and program requirements.
- It promotes a vision that simulation and modeling can be used effectively during the design of tooling in order to supplement in the decision-making process for engineers and program managers. Thus far in the industry, there has been limited work on the use of simulation for tooling design and it has been mainly based on experience from subject matter experts (SMEs) but this book intends to change that by introducing it to students that aspire to become engineers working in the manufacturing and design world as well as engineers in companies alike. Several detailed examples of simulation are provided to aid the reader in creating similar approach for future work.

Why this book?

One can always make an engineering judgment when it comes to design and analysis but it is extremely hard to know its impact on the production and only when early stage manufacturing starts will it be visible causing program delays and huge cost impact. Without prior knowledge on how a specific part or tool might behave, the only alternative will to rely on testing but that has also consequences in terms of schedule and cost that is not always available. Given the lack of books in the literature on tooling for composite structures focusing on aerospace parts this book is intended to fill that gap and help the industry to achieve better tools and a reference to guide for the design process. It also offers simulation examples that can be used to supplement some of the decisions and aid in creating substantiation for the decisions made.

Who is this book for?

The book targets undergraduate and graduate students in the field of manufacturing, design, structures, and materials in both aerospace and mechanical engineering. It is also intended for professional engineers in the field of tool design and composite manufacturing to be used as reference and guide. These types of tools are mostly used in the aerospace and automotive industry for part fabrication. They are also used in marine and sports applications. The book assumes that the reader has limited knowledge in composite materials and tries to provide a comprehensive explanation of the different aspects as the chapters' progress. The expectation is that the reader can fully understand the details of tooling and manufacturing of composite materials after the completion of the book. It will also shed light on how to use simulation and modeling to guide the design and offer better substantiation for the decision making process.

How to read this book

This book is designed as to provide a comprehensive explanation of the different topics involved in tooling for composite materials. The reader is encouraged to read the book starting from Chapter 1 and on until the end in a stepwise fashion. This will provide the most complete picture of all aspects. If the reader is experienced and would like to learn about specific topics offered in the book, the reader is encouraged to simply read the chapter of interest and all acronyms are fully defined to easily refer to them as needed.

Readers new to the field of tooling and composite manufacturing can obtain an overview by reading Chapter 1: Introduction to offer an overview of the topic. This will set the stage for the remaining of the chapters in the book.

Notes for instructors

This book can be used in an introductory course in tooling for composite materials and manufacturing or as a supplementary text in an undergraduate or graduate aerospace or mechanical engineering course focused on manufacturing and design. In particular, the instructor should start by motivating the topic using the content in Chapter 1 to offer the students a high-level overview of why this topic is of interest and the importance of the details introduced. Chapter 2 provides a high-level overview of composite materials. The instructor can skip this chapter if the students had already taken a more in-depth class but it's always good to refresh student's memory especially as the focus is on the materials aspect. Chapter 3 gets into the details of tooling as it pertains to materials and processing. This is an essential chapter and the instructor is encouraged to discuss it in detail. Chapter 4 will provide the students with an understanding of design requirements related to composite tooling for both lamination and assembly. Chapter 5 will go over the manufacturing details of composite materials and discuss several different methods and tooling approaches. This is followed by how to operate the tools in a production environment discussed in Chapter 6. Modeling and simulation is presented in Chapter 7. This is considered as a more advanced topic and the instructor should gage the students' background in this topic before he pursues the discussion. It is recommended that students are familiar with finite element analysis and some computed aided design (CAD) software. If not, the instructor is encouraged to discuss those topics briefly before proceeding to ensure that students get the most out of this chapter. Final thoughts and future work is discussed in Chapter 8.

List of Acronyms

ABS	acrylonitrile-butadiene-styrene
AF	assembly fixture
AFP	automatic fiber placement
AM	additive manufacturing
ATL	automatic tape layup
AWS	American Welding Society
BMC	bulk molding compound
BOM	bill of materials
BTP	build-to-print
BUPs	build-ups
CAD	computer aided design
CAM	computer-aided manufacturing
CDR	critical design review
CF	check fixture
CFD	computational fluid dynamics
CFRP	carbon fiber reinforced plastics
CHILE	cure hardening instantaneous linear elastic
CMM	coordinate measuring machine
COTS	commercial off-the-shelf
CTE	coefficient of thermal expansion
DFM	design for manufacturing
DMA	dynamic mechanical analysis
DOC	degree of cure
DSC	differential scanning calorimetry
EMR	external mold release agent
EOL	end of laminate
EOP	edge of part
EOP	end of part
ETD	engineering tool definition package
FDM	fused deposition modeling
FOD	foreign object damage
GSE	ground support equipment
HD-FOS	high definition fiber optic sensing
HF	handling fixture
Hg	mercury
HTC	heat transfer coefficient
ICD	interface control document
IML	inner mold line
IMR	internal mold release agent

INFX	inspection fixture
ITAR	International Traffic in Arms Regulation
LM	lamination mold
MLFX	mill or machining fixture
MM	master mold
MPS	manufacturing process specification
MSL	mean sea level
NDI	nondestructive inspection
NDT	nondestructive technique
OEM	original equipment manufacturer
OHME	overhead mechanical equipment
OML	outer mold line
OOA	out-of-autoclave
OTP	optical tool point
PCD	polycrystalline diamond
PDR	preliminary design review
QA	quality assurance
R&D	research and development
RMS	root mean squared
RTM	resin transfer molding
SCFM	self-consistent field micromechanics
SHF	special handling fixture
SMC	sheet molding compound,
SME	subject matter experts
SQRTM	same qualified resin transfer molding
ST	specialized tooling
TC	thermocouples
TD	trim and drill fixture
Tg	transition temperature
TGA	thermogravimetric analysis
TMA	thermal mechanical analysis
TOF	time of flight
TPC	thermoplastic composites
UHF	universal holding fixture
UV	ultraviolet
VCP	VERICUT composite programming

Chapter 1

Introduction

After more than 4 decades of using carbon fiber reinforced plastics (CFRP) (also referred to as composites for short) to build primary structures in the aerospace, automobile, civil, and ship industries; the industry is very much aware of the advantages they offer when it comes to airframe design such as: light weight, tailorability, improved fatigue performance compared to metallic materials, and much more. But utilizing composites to build parts did not transpire without setbacks [1–3], rather, a tremendous amount of engineering, cost, and time went into creating the right design, fabrication process in order to ensure safety, and improved performance which was the ultimate objective of utilizing these materials.

The high-specific strength and stiffness of composite materials make them an attractive candidate to replace metallic materials in order to reduce weight. The first composite components on commercial aircraft were designed and built as part of the NASA Aircraft Energy Efficiency (ACEE) program that entered service during the 1972–86 timeframe [4]. Several companies during that time participated in the program, which included Boeing Commercial Airplanes, Douglas Aircraft Company, and Lockheed Corporation. The objective was to get practical experience with composite components and to compare the long-term durability of flight components to data obtained from an environmental-exposure ground test program.

By January 1987, 350 composite components had entered into commercial airline flight service. Airbus was the first manufacturer to make extensive use of composite structures on large transport commercial aircraft on its A310 aircraft [5]. The A320 was the first aircraft to go into production with an all-composite empennage. Also, about 13% of the weight of the wing on the A340 is composite materials. The Boeing B-777 made extensive use of composite materials for primary structure in the empennage, most control surfaces, engine cowlings, and the fuselage floor beams. About 10% of the structural weight was composite materials [6]. On the Boeing product, graphite/epoxy composite materials were used for most secondary structures and control surfaces. A toughened epoxy material system, Toray T800H/3900-2, was used for the larger, more heavily loaded components including the vertical fin torque box and horizontal stabilizer torque box components of the empennage.

Rotorcraft and general aviation airframes have also used composite materials to achieve performance goals. The V-22 tiltrotor aircraft designed by Bell and Boeing has a number of significant applications of composite materials

Tooling for Composite Aerospace Structures. http://dx.doi.org/10.1016/B978-0-12-819957-2.00001-8

where 41% of the airframe was made from composite materials. Bell and Boeing used an integrated product team approach to design the V-22 airframe [7]. The approach is credited with saving about 13% of the structural weight, reducing costs by 22%, and reducing part count by about 35%.

When it comes to military aircrafts, they have leveraged composite materials expensively in their airframes. The Lockheed Martin F-22 Raptor is approximately 39% titanium, 16% aluminum, 6% steel, 24% thermoset composite materials, 1% thermoplastic composite materials, and 14% other material systems [8]. The wing skins are made of monolithic graphite/bismaleimide materials. The Lockheed Martin F-35 on the other hand had 35% of the airframe made from composites almost exclusively in skin applications [9]. The Northrop Grumman B-2 is constructed of mostly all composite materials [10]. Development of the B-2 began in the late 1970s. The first flight test of the B-2 was on July 17, 1989. The wing is mostly graphite/epoxy material with honeycomb skins and internal structure. The fuselage also makes extensive use of composite materials. The C-17 uses about 8% composite materials, mostly in secondary structure and control surfaces. In 1994, McDonnell Douglas attempted to redesign the horizontal tail using composite materials. The tail was redesigned using AS4 fiber in an epoxy resin for a 20% weight savings, 90% part reduction, 80% fastener reduction, and 50% acquisition cost reduction. The Boeing 787 uses composites materials in almost 50% of the airframe and was a leap in terms of the extensive use of composites for commercial airframes [11]. It led to a dramatic level of innovation in composite materials in terms of fabrication, analysis, design, tooling, handling, and many other disciplines including universities and research institutes. The most recent commercial airliner to use extensive composite materials is the A350 with about 53% of the structural weight [12]. Fig. 1.1 shows an example of different aircrafts and the

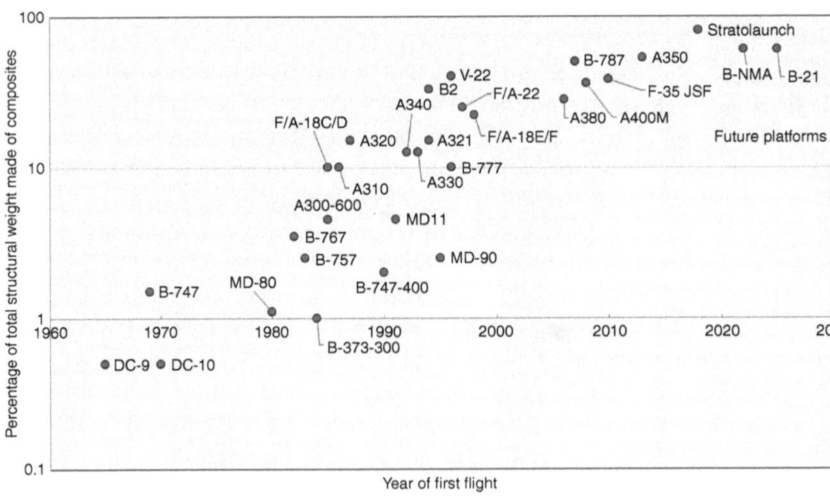

FIGURE 1.1 Product applications utilizing composite materials.

FIGURE 1.2 Airplanes that have majority of their airframe made from composite materials.

level of composite materials used on each. As observed, over the past 10 years an expensive amount of the airplanes have used composite materials for majority of the airframe structure leading the industry and supply chain to expand in order to meet the needs.

Some of those products are shown in Fig. 1.2. They range from commercial and military aircrafts, unnamed aerial vehicles, and experimental aircrafts among others.

But at this point you might be asking yourself especially if you have never worked with composite materials what are they? For the most part, when people in the aerospace industry refer to composite materials or composites for short

FIGURE 1.3 Composite materials (A) wet layup (B) plain weave fabric prepreg (C) unidirectional tape prepreg.

they are referring to carbon fibers that are embedded into an epoxy resin system as the matrix material holding them together. The material comes in the form of thin plies with different forms (e.g., unidirectional, plain weave). The material can also come in the form of prepreg referring to the product that has the matrix material pre-impregnated into the fibers or can be impregnated in a shop floor directly which is referred to as wet layup. Some of these forms are shown in Fig. 1.3. Additional discussion on composite materials will be provided in Chapter 2. In order for the material to be used it will need to be laid up in a tool known as lamination mold (throughout this book we use the word tool to refer to lamination molds) in order to take the shape of the intended design as shown in Fig. 1.4.

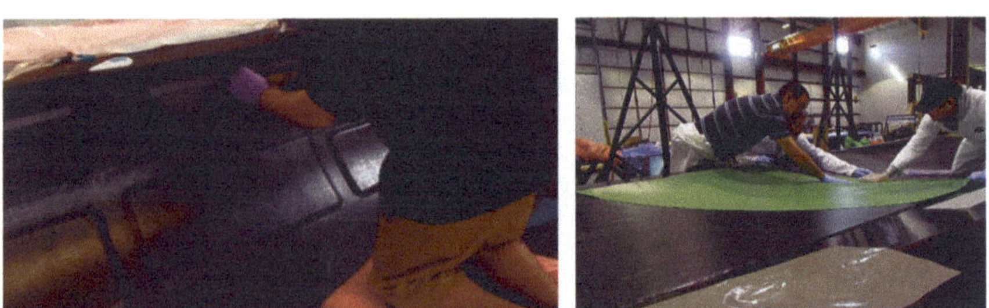

FIGURE 1.4 Typical composite lamination process on a tool.

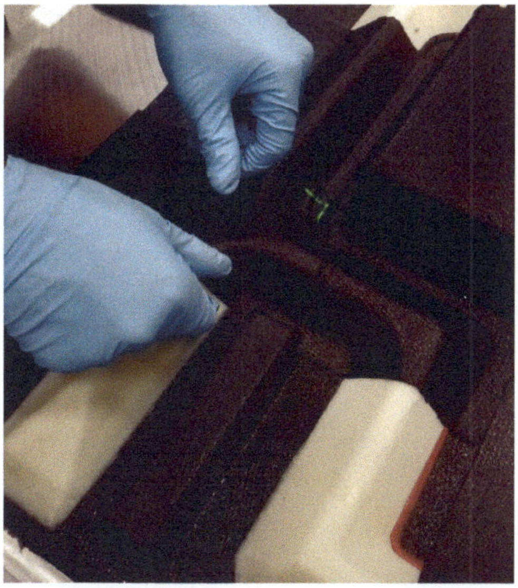

FIGURE 1.5 Part with complex shape being laminated requiring additional skill to prevent defects.

The lamination process occurs by taking the different material forms and cutting it into pieces and laying it up on a tool. As you can imagine it requires a large amount of skill to ensure that the layers are located appropriately along the part especially when the part is designed in a fashion that requires many joggles and sharp corners as shown in Fig. 1.5. A great deal of detail and methodical planning is typically required to make sure that the part is laminated to prevent any anomalies on the part post cure such as wrinkles and porosity that are known to occur frequently in thermoset materials. These types of defects are shown in Fig. 1.6. Depending on the location of where they exist on the part and type of structure (e.g., primary, secondary, tertiary) they might cause the rejection of the part and possibly scrapping it, which is costly to any program. There have been many proposals on ways to reduce the number of defects by either improving the lamination process all together or improving the design of parts to aid in the manufacturing process all of which will be discussed later in this book.

These tools such as that shown in Fig. 1.7 are a vital part of the composite design phase that are typically underestimated or overlooked given the lack of understanding among the engineering community regarding their importance in obtaining a final product that satisfies all requirements. A tool that is not designed correctly can cause the part to fail and be scrapped causing tremendous delays and quality concerns in production. This is why it is important during the preliminary design phase to evaluate tooling options and ensure that the appropriate level of trade studies have been made prior to putting an order for procurement which has a long-lead time and high cost and any mistake will

FIGURE 1.6 Anomalies observed on a part during lamination and post cure.

FIGURE 1.7 Example of a tool used for composite lamination.

cause long delays. These tools can be made from any material but the industry has consolidated its use to a smaller subset of materials for majority of tools depending on the application.

Some of the most common materials used to fabricate tools for composite material lamination are listed in Table 1.1.

Fiberglass, high-density foams, and tooling boards are candidates when considering parts that need to be cured at room temperature or prototype tooling. The dimensional accuracy will have larger variation compared to other material systems. However, when it comes to the fabrication of high-quality parts that require exceptional accuracy, other types of materials will need to be considered. The cost and complexity of the tooling is relative with the performance required and the number of production life cycles that we need the

TABLE 1.1 Material options used for composite lamination tool fabrication.

Material	Thermal conductivity (BTU/(hr ft °F))	Density (lb/ft^3)	Material cost	Repair process	Production cost
Aluminum	137	175	Med	Easy	High
Steel	25	480	Low	Hard	Med-high
Invar	7.5	508	High	Hard	High
Titanium	12	325	High	Hard	High
Ceramic	5.2	156	Low	Med	High
Carbon Fiber/ epoxy	3.0	93	High	Med	Low-High
Graphite	57	108	Low	Med	High
Fiber glass/ epoxy	2.5	118	Low	Med	Med-Low
Nickel	51	555	Med	Med	High
Carbon foam	11.5	17	High	Easy	High
Tooling board	0.15	55	Med-High	Easy	Low
Wood	0.03	31	Low	Easy	Med-Low

tool to survive, which can be in the hundreds of cycles. The tooling martial in which high performance composite parts are made can be carbon fiber/epoxy, monolithic graphite, ceramics, or metals, which are typically aluminum or steel.

Ceramic and metallic toolings are relatively heavy and expensive as compared to some of the composite material counterparts. They can range from hundreds to thousands of pounds in weight compared to ten to hundreds of pounds typically for composites. Example of different lamination tooling made from different materials is shown in Fig. 1.8. Invar is considered to be the most expensive of all materials but that also depends on the tool shape and size. In addition, few companies have the equipment necessary to cut and polish metal tools so they often require the services of a tooling specialist when any damage occurs to them.

Composite tooling on the other hand can be constructed more easily compared to invar, and since they are made from the same material as the part they can be made in-house by many manufacturers depending on the capacity and amount of risk the company is willing to take. One of the drawbacks that come with this tooling is the vulnerability to wear and damage but are easier to repair as compared to invar. Keep in mind that the way to manufacture metallic tooling is by either machining or through a bump forming process for invar specifically as shown in Fig. 1.9. The fabrication of composite tooling will need to have

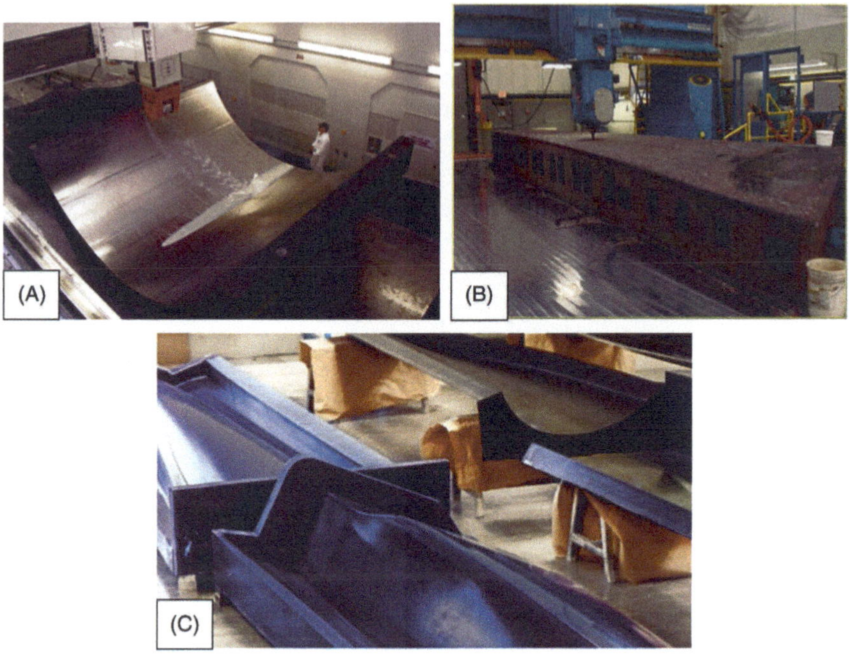

FIGURE 1.8 Different tooling materials used (A) invar tooling (B) composite tooling (C) high-temperature foam tooling [13–15].

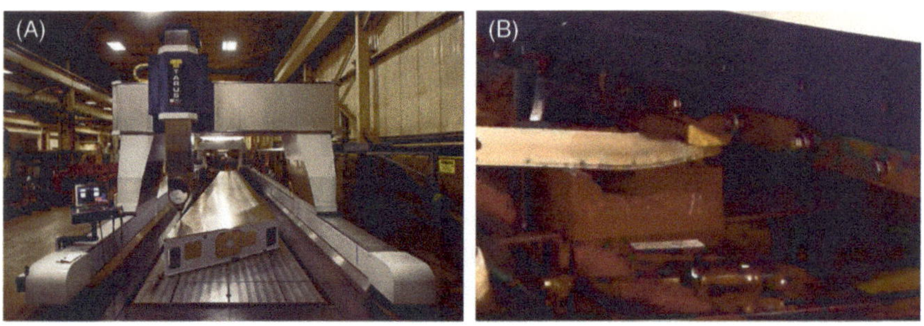

FIGURE 1.9 Metal tool fabrication (A) aluminum machining (B) invar bump forming [16–17].

a master mold that is built as a secondary tool in order to layup the laminates similar to how you will build a composite part, this is followed by including an egg crate piece as needed for handling purposes. Details of the manufacturing of composite tools are shown in Fig. 1.10.

One of the key issues when it comes to tooling is the phenomenon of coefficient of thermal expansion (CTE) mismatch between the tool and the part during cure. This mismatch can cause adverse effect on the part such as warpage due to residual stresses accumulated during cure. More details on this topic will be discussed in later chapters.

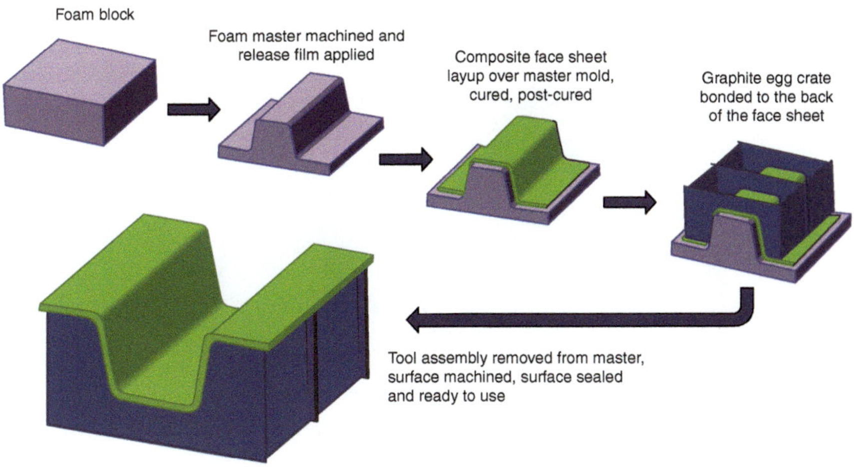

FIGURE 1.10 Composite tooling fabrication steps.

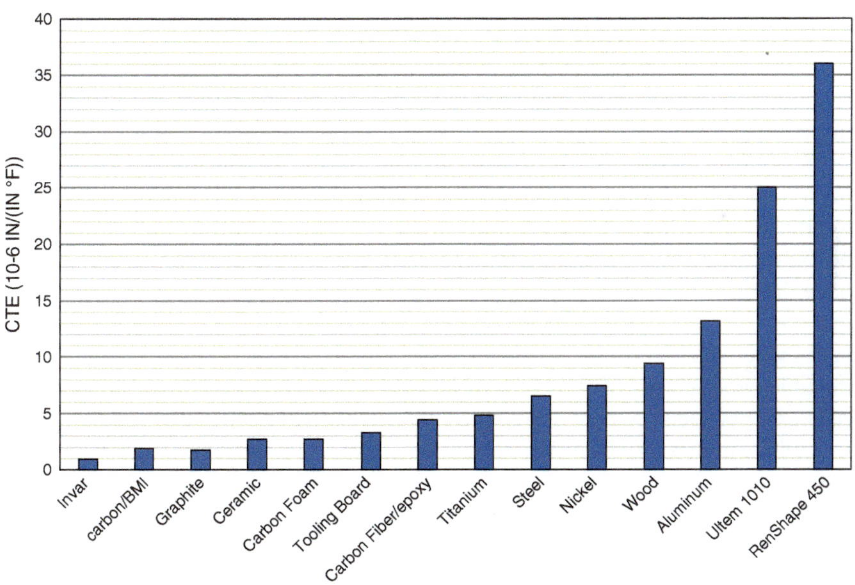

FIGURE 1.11 CTE variation of different materials used for tooling.

Most metallic tooling materials and composites have mismatched CTE as shown in Fig. 1.11. Although steel and aluminum are common choices to build tools since they are less expensive and usually involve shorter production lead times; during the cure cycle, the CTE mismatch is often too extreme for use with close-tolerance composite parts. Only the higher-priced metal alloys, such

FIGURE 1.12 HyVarC tooling with invar base and composite face sheet.

as invar, offer closer CTE matches that can be an option to consider keeping in mind the cost and lead-time differences.

Composite tools made from tooling prepregs have a CTE close to the part CTE, helping the part maintain dimensional accuracy during cure. Shrinkage and thermal expansion of the tool and part will be very similar making them more attractive as an option.

Since invar offers a CTE very close to that of carbon fiber composites it has been the typical choice for parts that must be manufactured to extremely tight tolerances. But due to the size and weight of those tools, the handling process becomes much more challenging especially for smaller companies that do not have the resources such as large spaces and cranes. There is a desire among the industry for new tooling materials that can withstand thousands of autoclave cure cycles like invar but are cheaper and lighter. To date there has been a tremendous amount of work by many companies to come up with such new materials and products. Ascent aerospace offers a hybrid version that uses an invar tool with a composite face sheet as shown in Fig. 1.12. They promote it as being 50% lighter than a traditional invar tool and possess a 20% shorter lead-time than a traditional composite tooling solution, which is a very attractive alternative. Another advantage this type of tooling offers is the ability to machine the face if for example the part mold design changed after the tooling has been fabricated. The alternative will be to redesign a tool and fabricate a new one and that can be very costly to a program.

Other options include the use of bismaleimide (BMI) materials that offer high durability but are more expensive due to its processing. An example of such tool is the HexTool M61 offered by Hexcel as shown in Fig. 1.13. It was proven to be dimensionally stable with the ability to hold vacuum integrity before and after machining and after more than 500-part cure cycles. They are shown to be lighter in weight as compared to invar and lower cost.

FIGURE 1.13 HexTool® M61 tooling fabricated from Hexcel's BMI resin.

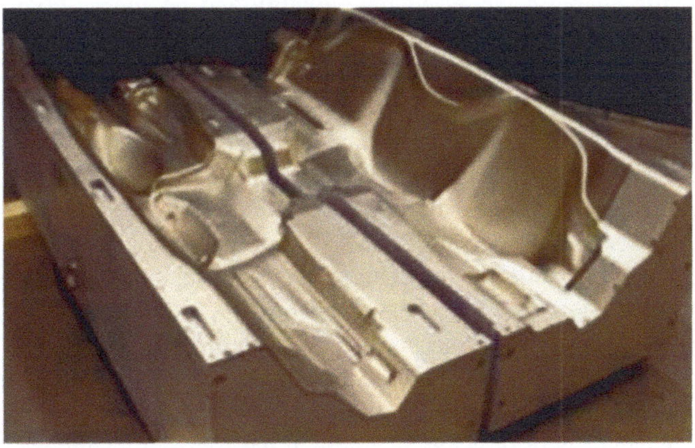

FIGURE 1.14 Nickel Vapor Deposition (NVD) shell tooling.

Another alternate to traditional metals for tool fabrication is the use of nickel vapor deposition (NVD) process to produce relatively thin nickel-shell tool faces supported on a backing structure as shown in Fig. 1.14. The nickel tool surface can achieve high-dimensional accuracy, and offers low CTE, long life, and given its conformal design, it weighs less and facilitates faster mold heating and cooling. This type of tooling is used more often for the fabrication of automotive parts and not as much when it comes to aerospace composite parts.

Lately the use of additive manufacturing (3D printing) specifically fused deposition modeling (FDM) is becoming the technology of choice for rapid tooling production. Examples of such tools are shown in Fig. 1.15. There are many companies that have been working on advancing the use of 3D printing technology for this purpose. As of now it has been mainly used for 350 °F cure temperatures as well as moderate-temperature (less than 325 °F) production tooling, low-volume that does not require many cycles, and other repair tools. Relative to traditional tooling materials and methods, FDM offers significant advantages in terms of lead-time, tool cost, and simplification of tool design, fabrication

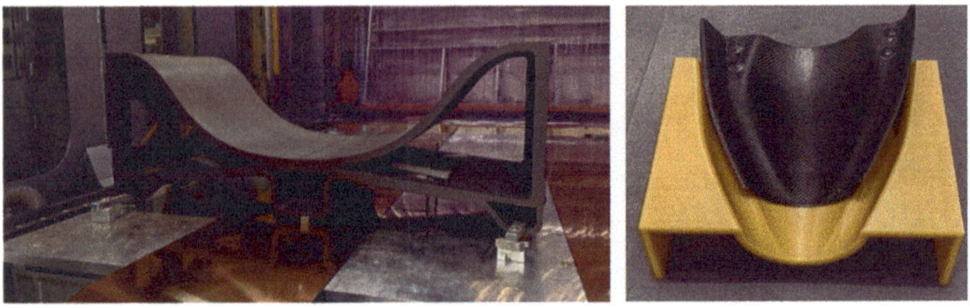

FIGURE 1.15 Additive manufacturing used to build lamination tools [18–19].

and use, while enabling increased functionality and geometric complexity. In Chapter 8 we will be discussing the use of additive manufacturing tooling in more detail including materials, manufacturing, and operational aspects.

Another important aspect when it comes to the planning phase of any project is the cost of the tools as they are considered one of the largest non-recurring cost when it comes to airframe design and production of any project. We have briefly eluded too some of the material cost, but a comprehensive evaluation is required in order to consider all aspects. To produce a tool in a cost effective manner, you need to optimize the design, material usage, labor, and machining hours spent.

As for tool design, two primary approaches are usually considered, a female tool approach as shown in Fig. 1.16A, or a male tool as shown in Fig. 1.16B. Depending on which side of the part that needs the right size and surface finish, either option can be chosen. As an example, if we consider a wing we know that the out mold line (OML) is important as it will be the side exposed to the airplane exterior and requires a smooth finish to reduce drag. In that case, a female tool may be more appropriate compared to a male tooling approach. It is also possible to use a matched die tool, where both female and male tools are used. This is a good way to control the thickness, but it has high-tooling cost.

Other aspects of the part fabrication that need to be considered when selecting the tool is the environment where the tool will be operating in. For example, will the tool be exposed to high temperatures or pressure? What are the total number of parts expected to be produced on the tool? What are the final part

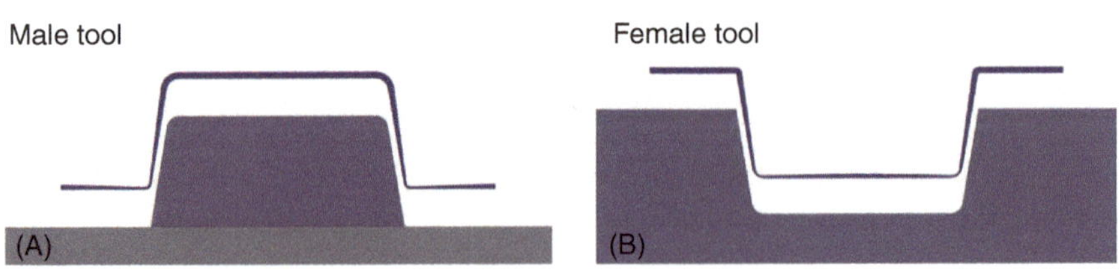

FIGURE 1.16 Different tooling approaches (A) male tool (B) female tool.

tolerance needed? All those questions have a big influence on both the production method and the material selection of the tool.

When it comes to parts with low-production quantity or that might be used as prototype parts, materials such as fiberglass, high-density foams, epoxy boards, or wood/plaster models have often been used. For these materials, the curing of the parts shall occur at ambient or low temperatures to ensure the survival of the tool for at least couple cycles. Moreover, to keep the cost low, dimensional accuracy to tight levels may not be necessary to achieve the objective and higher tolerances may be used. Additive manufacturing has also been used more often in recent years for tooling fabrication as an alternative to prototype tools.

High-rate production tools on the other hand are generally made of tough metals that can survive repeated cure cycles and maintain good surface finish and dimensional accuracy. The tools in which high-performance composite parts are formed can be made from carbon fiber/epoxy, monolithic graphite, ceramics or metals. In all cases, each material offers unique capabilities and drawbacks.

There are two general types of tooling—hard tooling and soft tooling. Hard tooling involves using metallic materials such as steel, aluminum, or metal alloys like invar; while soft-tooling materials are typically composite materials such as fiberglass, high-density foam, machinable epoxy boards, or wood/plaster models. Traditionally, hard tooling has been the primary option for most aerospace applications given it is durable quality, good surface finish, and ability to withstand high-production cycles (e.g., up to 1500 autoclave cycles for steel tools). Moreover, metals generally have low CTEs, which works well when producing components that also have a low CTE, require repeated high-temperature cycling, or demanding tolerances. By matching the CTE values of the tooling and production material, they will both expand and contract at the same rates when exposed to varying curing temperatures—resulting in high-quality parts with precise dimensional tolerances.

When considering manufacturing composite parts, soft tooling options are often a better choice to reduce CTE mismatch and maintain dimensional accuracy during a cure. Advantages that soft tooling offers compared to hard tooling, is the ease of machining into complex shapes and rapid rework capability. The soft tooling raw material, as well as the process of machining it, also comes with a lower price tag, and is easier to maneuver with its lighter weight. These benefits are why more and more manufacturers are moving to lighter-weight soft tooling for prototyping and other time-sensitive projects, creating tools with large or complex designs, production runs with low-part volumes, or when low-cost solutions are required.

Given the complexity of fabricating tooling for the aerospace industry there are a finite number of companies that produce most of the tools. A list of those companies is shown in Table 1.2, note that this is not a comprehensive list but a subset of companies that the author has direct experience with. This list also excludes the actual original equipment manufacturer (OEM) that sometimes

TABLE 1.2 Companies that manufacture tooling for the aerospace industry.

Company	Main production location	Revenue	Number employees	Year founded	Materials used for tool fabrication	Industry focus
Janicki Industries	Hamilton, Washington	$100–$500 million	600	1993	• Block Foam • Composites • Invar • Steel • Aluminum	• Aerospace • Marine • Energy • Transportation • Architecture
EireComposites Teo	Galway, Ireland	NA	70	1997	• Invar • Steel • Aluminum	• Aerospace • Energy
Quickstep Holdings Ltd.	Victoria, Australia	$60 million	220	2001	• Composites	• Aerospace • Automotive
Weber Manufacturing Technologies Inc.	Midland, Ontario	$28 million	200	1962	• Invar • Steel • Aluminum • Nickel Shells	• Automotive • Aerospace • Building
North Coast Tool and Mold Corp.	Cleveland, Ohio	NA	75	1976	• Invar • Steel • Aluminum	• Aerospace
Touchstone Research Laboratory	Triadelphia, West Virginia	$5–$10 million	23	1980	• Carbon foam • Composites	• Aerospace • Automotive • Chemical • Construction

Models & Tools	Shelby Charter Township, Michigan	NA	51–100	1974	• Composites • Invar	• Aerospace • Communications
Advanced Integration Technology	Grand Prairie, Texas	$10 million	900	1992	• Composite	• Aerospace
Ascent Aerospace	Macomb Township, Michigan	$386 million	1,100	2014	• Invar • Aluminum • Steel • Composites	• Aerospace
Futuramic	Warren, Michigan	$43 million	145	1955	• Aluminum • Stainless Steel • Invar	• Aerospace
Visioneering	Auburn Hills, Michigan	$21.6 million	101-250	1953	• Invar	• Aerospace
Vector	Winfield, Kansas	$0.746 million	51–200	2013	• Aluminum • Steel • Invar	• Aerospace
Sawyer	Fort Worth, Texas	>$1.0 million	11–50 employees	1992	• Composites • Aluminum	• Mechanical • Industrial

builds their own tools internally. All the data in the table are based on open source information that can be found online.

Another type of tooling that manufacturers require during the fabrication of any product are assembly and ground support equipment (GSE) tooling. These are the tooling that operate at room temperature and have different type of requirements compared to lamination tooling that have been the focus of the discussion thus far. Figs. 1.17 and 1.18 show an example of assembly and GSE tooling used in production, respectively. Since these types of tools do not experience high temperatures, the issue of CTE mismatch is less of a concern especially if the assembly process occurs inside an enclosed building. Rather, one of the biggest requirements that need to be addressed in this case are tolerances and strength.

The tolerances on these tools need to ensure that when bringing the parts together for the mating process they allow the parts to fit in a seamless manner

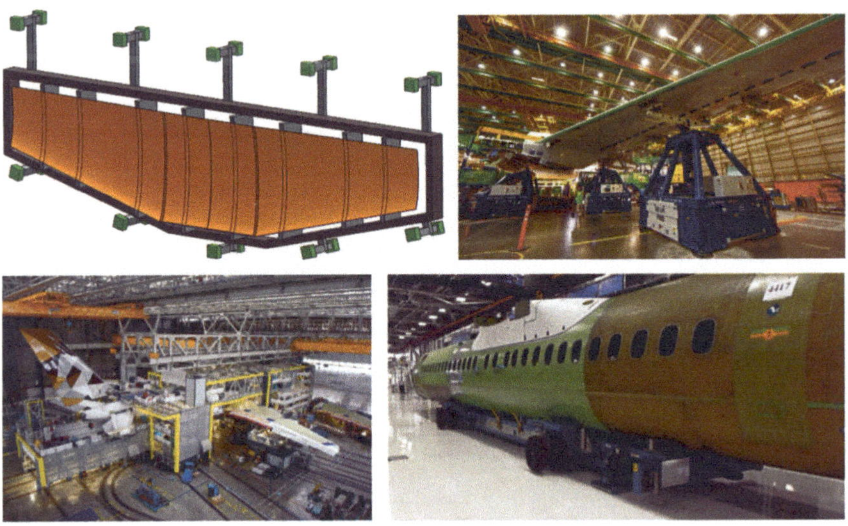

FIGURE 1.17 Assembly tooling used for airplane production.

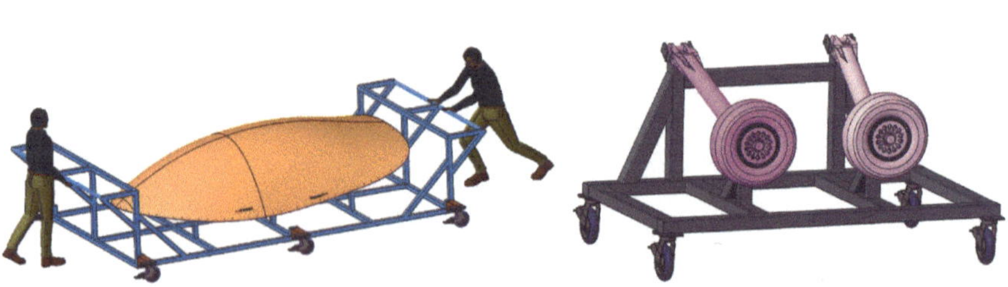

FIGURE 1.18 Example of ground support equipment (GSE) tools.

FIGURE 1.19 Match drilling process in aerospace components.

without exerting forces on the part causing them to fail or get damaged. Many tools are designed in such a way that enable the tool to be modified to tailor to the mating part, but these added features can be costly and add complexity to the tool with adverse consequences.

In addition, the tools need to be designed to withstand a level of stress by the mating parts without causing damage or failure. A factor of safety between 3 and 5 need to be included during the design phase of these tools as they are typically ordered for a program once and are expected to remain until the end of the program. Without having appropriate tooling for assembly and handling, many failures can occur during production causing huge amount of nonconformance and delays to the program sacrificing quality.

As an example, say that a program wanted to cut cost and delay the order of assembly tools needed for production. And instead of getting a tool that would have allowed two parts to mate and match drill as shown in Fig. 1.19, they decided to hand drill each part separately. In the hand drilling case, you will find it is very hard to adhere to tolerance typically used for holes in the aerospace industry of +/− 0.03 inches, which ultimately generates mismatches between the parts and the inability to assemble. This requires the need for an extensive repair and in some cases scrapping the part. Fig. 1.20 shows a part that has been hand drilled and the deviation measured between the "should be location" based on the CAD and the actual part post drilling. The deviation measured was around 0.135 inches and should have been 0.03 inches causing the need to repair the part by potting the holes and redrilling.

With the content discussed thus far the reader should have received a good grasp of the topic of tooling and the importance of why we need to dedicate an

FIGURE 1.20 Part that was hand drilled and metrology scan between CAD model and the part showing a 0.135-inch average deviation which is outside the allowed tolerances.

entire book to discuss this topic. In the next several chapters will dive deeper in each of the topics highlighted here and offer a comprehensive overview of tooling and ways to use it weather you are an engineer working in the field of design or manufacturing or a student new to this subject.

Chapter questions

1. Define tooling in the context of aerospace applications?
2. Name five materials used for composite tooling lamination?
3. Name three differences between lamination tooling and assembly tooling?
4. What thermal material property has a large impact on tooling performance?
5. What is the typical temperature that assembly tooling operates in?
6. What is the typical temperature that composite lamination tooling operate in?
7. What is the recommended factor of safety to design assembly tooling?
8. Provide one adverse effect of having large CTE mismatch between tool and part?
9. Which material used for typical composite lamination tools has the highest CTE?
10. Does having the incorrect assembly tooling have any negative impact on production? Explain?

References

[1] Burgoyne C. Fiber reinforced polymers- strengths, weaknesses, opportunities and threats. In: FRPRCS-9, 13-15 July. Sydney, Australia, 2009.

[2] https://www.technologyreview.com/s/409929/boeings-composite-problem/.

[3] http://www.lemauricien.com/article/airbus-versus-boeing-composite-materials-sky-s-limit.

[4] Dow MB. The ACEE Program and Basic Composites Research at Langley Research Center (1975 to 1986), Summary and Bibliography. Hampton, VA: NASA Reference Publication 1177. National Aeronautics and Space Administration, Langley Research Center, 1987.

[5] Pora J. Composite materials in the airbus A380 - from history to future. Society of Manufacturing Engineers, 2000.

[6] Smith B. The Boeing 777 Advanced Materials and Processes, 2000.

[7] Tenney DR, Davis Jr JG, Pipes B, Johnston N. NASA Composite Materials Development: Lessons Learned and Future Challenges. NATO RTO AVT-164 Workshop on Support of Composite Systems, 2009.

[8] Harris CE, Starnes JH, Jr, Shuart MJ. An assessment of the state-of-the-art in the design and manufacturing of large composite structures for aerospace vehicles. Hampton, VA: NASA/ TM-200 1-2 10844, National Aeronautics and Space Administration, 2001.

[9] Composites World. Skinning the F-35 fighter. compositesworld.com. Retrieved 24 November 2016.

[10] Boyne, Walter J. Air Warfare: an International Encyclopedia: A-L, Santa Barbara, California: ABC-CLIO, ISBN 978-1-57607-345-2., 2002.

[11] Safety, A. Status of FAA's actions to oversee the safety of composite airplanes, US Government Accountability Office report: GAO-11-849, 2011.

[12] Roth Y. Liquid composite moulding – enabler for the automated production of CFRP aircraft components, Eccm16-16th European Conference on Composite Materials, Seville, Spain, 22-26 June 2014.

[13] https://www.janicki.com/products/metal-tooling.

[14] http://compositesmanufacturingmagazine.com/2015/11/proper-tooling-design-matters.

[15] https://www.generalplastics.com/products/fr-4700.

[16] https://bernardandcompany.wordpress.com/2010/02/01/siemens-coast-composites-achieves-higher-accuracies-on-invar-tooling-for-aerospace-structures.

[17] https://www.youtube.com/watch?v=trHQX3KkeNI.

[18] https://www.stratasys.com/tooling/composite-tooling.

[19] https://www.additivemanufacturing.media/blog/post/webinar-additive-manufactured-tooling-for-composites.

Chapter 2

Composite materials

Chapter outline

The following chapter will provide additional details that pertain to composite materials. This content presented here is not a substitute for a class in composite materials rather a preliminary introduction for those that have not worked with this material before. The focus will be on the materials aspect and limited discussion is dedicated to the mechanics side.

A composite material can be defined as a combination of two or more materials that results in better properties than those of the individual components used alone. In contrast to metallic materials, each component retains its separate chemical, physical, and mechanical properties. The two constituents are reinforcement and matrix. The main advantages of composite materials are their high strength to stiffness ratio when compared with bulk materials, allowing for a weight reduction in the finished part. The reinforcing phase provides the strength and stiffness. In most cases, the reinforcement is harder, stronger, and stiffer than the matrix. The reinforcement is usually a fiber or a particulate depending on the application and needs. Particulate composites have dimensions that are approximately equal in all directions. They may be spherical, platelets, or any other regular or irregular geometry including carbon nanotubes. Particulate reinforced composites usually contain less reinforcement (40%–50% by volume) due to processing difficulties in dispersing the particulates into the matrix.

A fiber has an aspect ratio that is much higher as compared to smaller particulates which also impacts the overall performance of the composite material. Continuous-fiber composites normally have a preferred orientation, while discontinuous fibers generally have a random orientation. Examples of continuous reinforcements include unidirectional, woven fabrics, and helical winding while

Tooling for Composite Aerospace Structures. http://dx.doi.org/10.1016/B978-0-12-819957-2.00002-X

FIGURE 2.1 **Typical composite reinforcement (A) unidirectional (B) woven (C) winding (D) fiber mat (E) chopped fibers [1–5].**

examples of discontinuous reinforcements are chopped fibers and random mat as shown in Fig. 2.1. These composites are often made into laminates by stacking single sheets of continuous fibers in different orientations to obtain the desired strength and stiffness properties with fiber volumes as high as 60%–70% by volume. Fibers produce high-strength composites because of their small diameter that have an inverse relation with strength (the smaller the diameter of the fiber, the higher its strength) but often the cost increases as the diameter becomes smaller. In addition, smaller-diameter high-strength fibers have greater flexibility and are more flexible to fabrication processes such as weaving or forming over radii.

Typical fibers include glass, aramid, and carbon, which may be continuous or discontinuous. The continuous phase is the matrix, which is a polymer, metal, or ceramic. Polymers have low strength and stiffness, metals have intermediate strength and stiffness but high ductility, and ceramics have high strength and stiffness but are brittle. The matrix performs several critical functions, including maintaining the fibers in the proper orientation and spacing and protecting them from abrasion and the environment. In polymer and metal matrix composites that forms a strong bond between the fiber and the matrix, the matrix transmits loads from the matrix to the fibers through shear loading at the interface. In ceramic matrix composites, the objective is often to increase the toughness rather than the strength and stiffness; therefore, a low interfacial strength bond is desirable. The type and quantity of the reinforcement determines the final properties.

Given the advantages of composite materials, they have been used extensively in many aerospace products where majority of the airframe for each of those platforms is made from a variant of different composite materials.

Even though there are many available resources that are devoted to discussing composites materials [6–11] we would like to shed some light on important

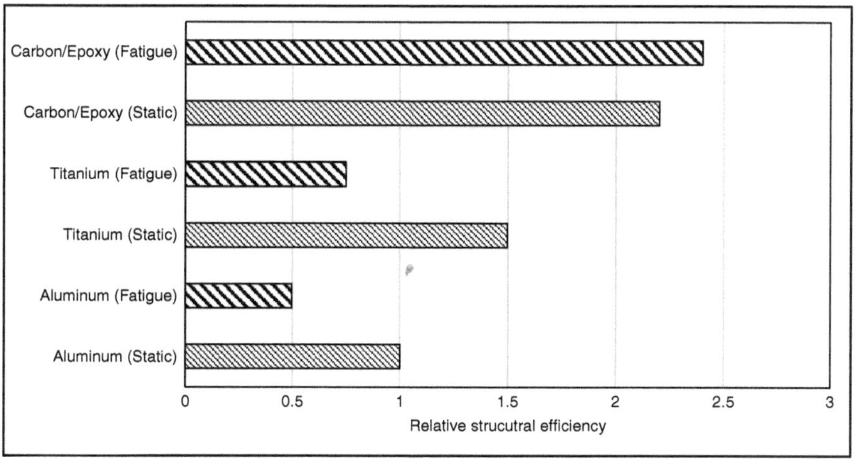

FIGURE 2.2 **Static and fatigue performance of several materials.**

aspects that the reader is required to know prior to reading the remaining chapters. This chapter is devoted to providing some basics of composites mainly carbon fiber/epoxy composites.

If you heard anything about composites thus far it is probably that they have high strength to weight ratio compared to metallic materials. This property by itself makes it an attractive option to be used where weight is important in applications such as aerospace products. Composites also have good fatigue performance, which is another property of interest to the aerospace field. Fig. 2.2 shows a comparison between aluminum, titanium, and composites when it comes to fatigue and strength performance and it is clear that composites have an advantage.

In addition to its properties, composites are fundamentally different in the way they are fabricated. Traditional techniques used for metals such as casting, and machining are not applicable. Composites on the other hand have many options of how they can be fabricated which depend on the form. Some of those options are summarized in Fig. 2.3. More details on this topic will be discussed in Chapter 5.

With this basic information at hand, let us dive into some nomenclature related to composites that will help the reader in understanding some of the terminology when dealing with this material.

1 Lamina

Consider the unidirectional fiber-reinforced composite ply also known as a lamina shown in Fig. 2.4. The coordinate system used to describe the ply is labeled the 1-2-3 axes. In this case, the 1-axis is defined to be parallel to the 0 degree fibers, the 2-axis is defined to lie perpendicular to the 0 degree fibers, and the 3-axis is defined to be normal to the plane of the plate. The 1-2-3 coordinate system is referred to as the principal material coordinate system. If the plate is loaded parallel to the fibers, the modulus of elasticity E_{11} approaches that of the

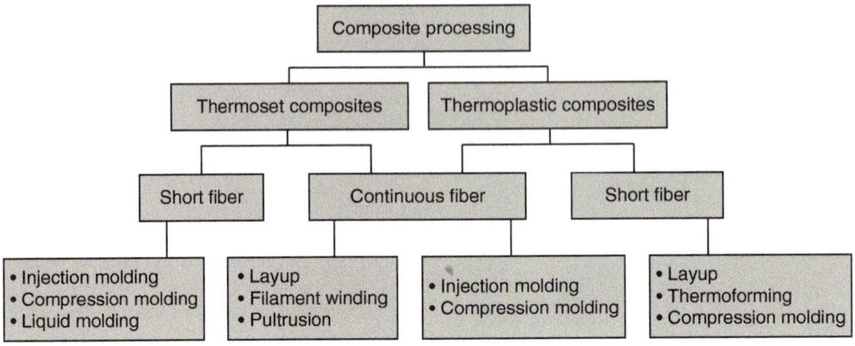

FIGURE 2.3 **Composite material processing.**

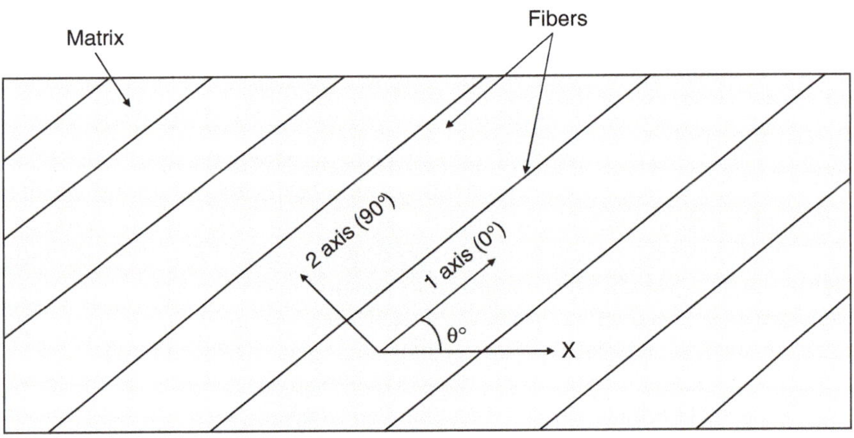

FIGURE 2.4 **Ply angle definition.**

fibers. If the plate is loaded perpendicular to the fibers in the two or 90 degree direction, the modulus E_{22} is much lower, approaching that of the relatively less stiff matrix. Since the modulus varies with the direction, the material is called anisotropic referring to properties that are direction dependent.

2 Laminate

When the plies are stacked at various angles, the layup is called a laminate. Continuous fiber composites are normally laminated materials as shown in Fig. 2.5 in which the individual layers, plies, or lamina are oriented in directions that will enhance the strength in the primary load direction. Unidirectional (0 degree) lamina are extremely strong and stiff in the 0 degree direction. However, they are very weak in the 90 degree direction because the load must be carried by the much weaker polymer matrix. While a high-strength fiber can have a tensile strength of 500 ksi or more, a typical polymer matrix normally has a tensile

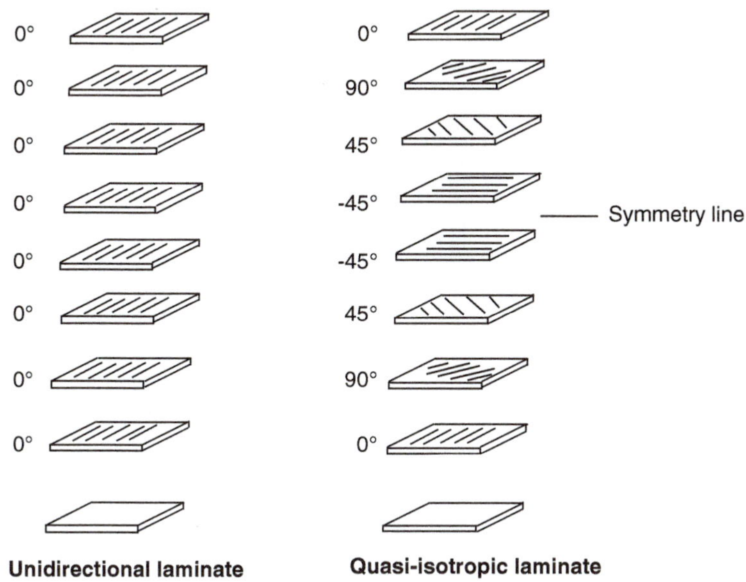

FIGURE 2.5 **Laminate stacking sequence.**

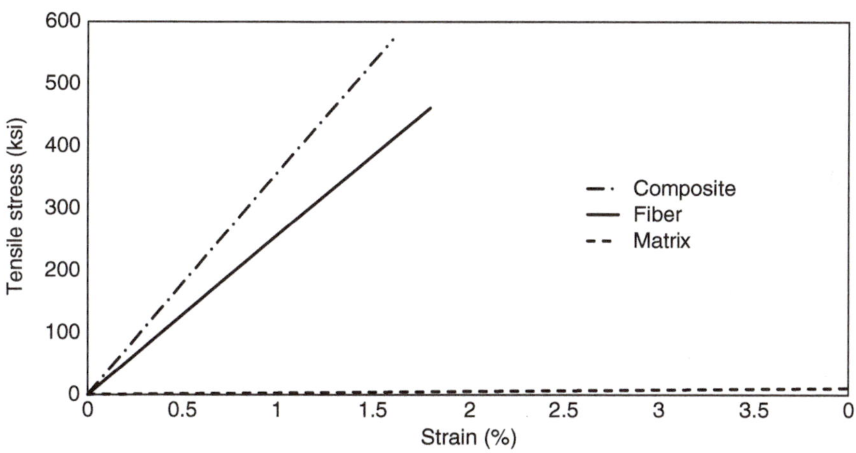

FIGURE 2.6 **Comparison of fiber, composite, and matrix properties.**

strength of only 5–10 ksi as shown in Fig. 2.6. The longitudinal tension and compression loads are carried by the fibers, while the matrix distributes the loads between the fibers in tension and stabilizes the fibers, preventing them from buckling in compression. The matrix is also the primary load carrier for interlaminar shear (i.e., shear between the layers) and transverse (90 degree) tension.

Once the layers of prepreg are laid-up on a tool they are cured as shown in Fig. 2.7. The layers are laid up in the required directions and to the correct

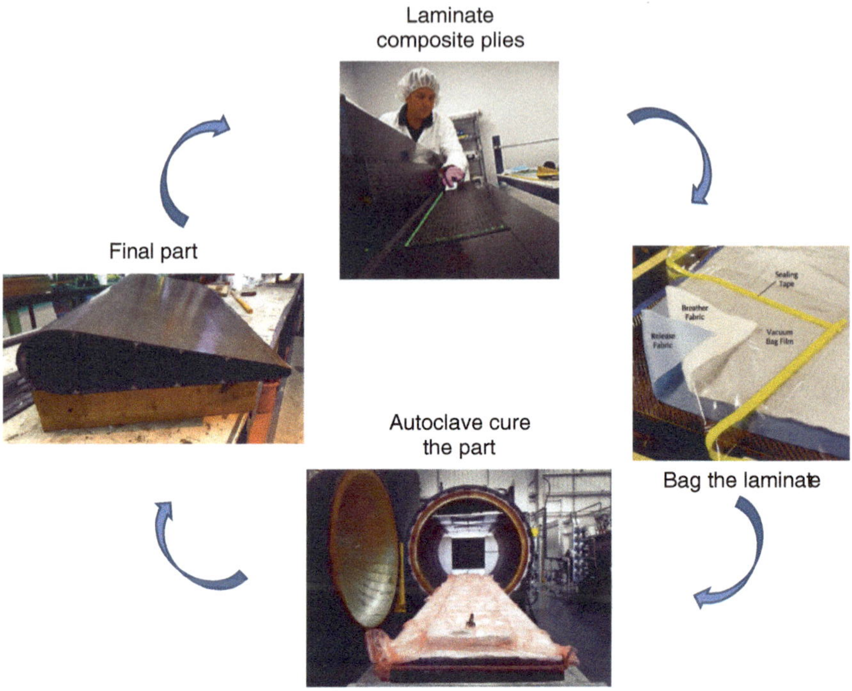

FIGURE 2.7 **Composite part lifecycle [12–15].**

thickness. A thin nylon vacuum bag is then placed over the lay-up and the air is evacuated to draw out the air between the plies. The bagged part is placed in an oven or an autoclave (a heated pressure vessel) and cured under the specified time, temperature, and pressure per a predefined specification. If oven curing is used, the maximum pressure that can be obtained is atmospheric (14.7 psia or less). An autoclave shown in Fig. 2.8 works on the principle of differential gas pressure. The vacuum bag is evacuated to remove the air and the autoclave supplies gas pressure to the part. The vessel contains a heating system with a blower to circulate the hot gas. An autoclave offers the advantage that much higher pressures (e.g., 100 psig) can be used resulting in better compaction, higher fiber volume percentages, and less voids and porosity.

3 Fundamental property relationships

When a unidirectional continuous fiber lamina or laminate is loaded in a direction parallel to its fibers (0 degree direction), the longitudinal modulus E_{11} can be estimated from its constituent properties by using what is known as the rule of mixtures:

$$E_{11} = E_f V_f + E_m V_m$$

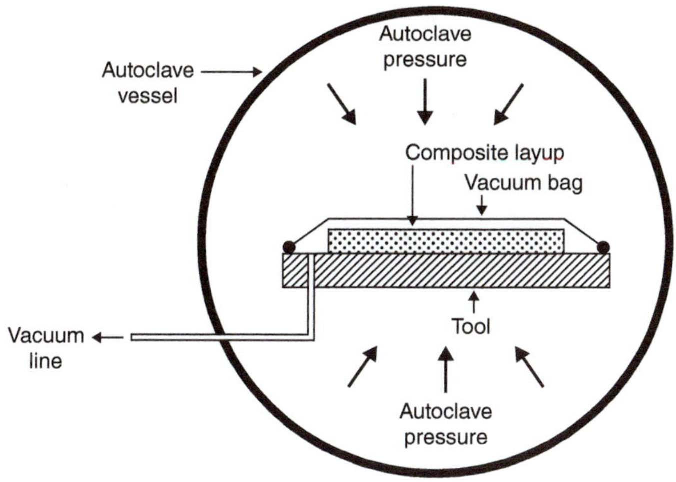

FIGURE 2.8 Autoclave principle.

where E_f is the fiber modulus, V_f is the fiber volume percentage, E_m is the matrix modulus, and V_m is the matrix volume percentage. The longitudinal tensile strength σ_{11} can also be estimated by the rule of mixtures:

$$\sigma_{11} = \sigma_f V_f + \sigma_m V_m$$

where σ_f and σ_m are the ultimate fiber and matrix strengths, respectively.

For more detailed information on how to conduct analysis to compute the effective material properties for composite laminates the reader is directed to [9] where extensive analysis on this topic is presented. For the purposes of this book, we will not go over any additional details on this manner.

4 Fabrication of composite prepreg

The method of how to fabricate composites as discussed earlier can be done in many ways but the most common method used to fabricate aerospace parts is using prepreg and hand layup on tools. Prepreg is simply a form of composite where the matrix and fiber are impregnated together to form rolls as shown in Fig. 2.9. The fabrication approach starts with fibers coming out of different spools and are deposited onto thin films of matrix material in such a way to achieve the desired fiber fraction and tack level via different compactions and heating cycles along the way. An example of such fabrication is shown in Fig. 2.10.

The advantage of using prepreg is that it offers consistency in the way parts are fabricated but given the complexity in the manufacturing approach this material form can be expensive compared to wet layup for example. Some additional comparisons are shown in Table 2.1.

Composite materials also have a finite life span specifically in the prepreg form. Generally speaking, they have a 12-month life span, if they are stored at

FIGURE 2.9 **Composite roll of unidirectional prepreg [16].**

FIGURE 2.10 **Prepreg line [17].**

TABLE 2.1 Comparison between wet layup and prepreg.

Property	Wet layup	Prepreg
Cost	Lower	Higher
Shelf life	Better	Worse
Storage	Better	Worse
Drapeability	Better	Worse
Tack	Worse	Better
Resin control	Worse	Better
Fiber volume control	Worse	Better
Part quality	Worse	Better

temperature 0 °F or 30 days at 70 °F. As time passes by the workability of the material becomes more difficult to deal with which is something that requires attention. There has been research on expanding the life span of the material but no clear direction in the industry has been agreed on and it is recommended to follow the manufacturers guidelines.

5 Types of fiber

Fibers have a very long axis compared to smaller particulates. They are usually circular or nearly circular and are significantly stronger in the longer direction, because they are normally made either by drawing or pulling during the manufacturing process, which orients the molecules so that tension loads on the fibers pull more against the molecular chains themselves than against a mere entanglement of chains. Due to the strength and stiffness of fibers, they are the predominantly used as reinforcements for advanced composites. Fibers may be continuous or discontinuous, depending on the application and manufacturing process.

5.1 Fiber terminology

Before examining the various types of fibers used for composite reinforcements, the major terminology used in fiber technology will be reviewed.

1. Fiber—a general term for a material that has a long axis, which is many times greater than its diameter. The term "aspect ratio" (fiber length divided by its diameter) is frequently used to describe short-fiber lengths. For fibers, aspect ratios are normally greater than 100.
2. Filament—the smallest unit of a fibrous material. For spun fibers, this is the unit formed by a single hole in the spinning process. It is synonymous with fiber.
3. End—a term used primarily for glass fibers, which refers to a group of filaments in long-parallel lengths.
4. Strand—another term associated with glass fibers that refers to a bundle or group of untwisted filaments. Continuous strand rovings provide good overall processing characteristics through fast wet-out (penetration of resin into the strand), and even tension and abrasion resistance during processing. They can be cut cleanly and dispersed evenly throughout the resin matrix during molding.
5. Tow—similar to a strand, tow is used for carbon and graphite fibers to describe the number of untwisted filaments produced at one time. Tow size is usually expressed as Xk. For example, a 12k tow contains 12,000 filaments.
6. Roving—a number of strands or tows collected into a parallel bundle without twisting. Rovings can be chopped into short fiber segments for sheet molding compound (SMC), bulk molding compound (BMC) or injection molding.

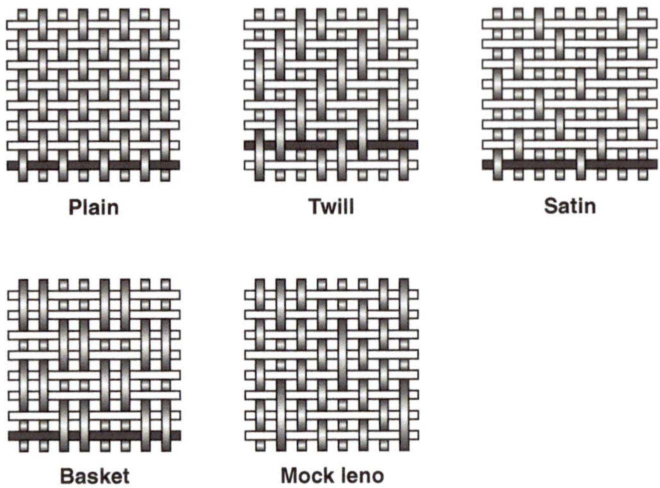

FIGURE 2.11 **Different fabric forms used in aerospace applications [18].**

7. Yarn—a number of strands or tows collected into a parallel bundle with twist. Twisting improves the handleability and makes processes such as weaving easier, but the twist also reduces the strength.
8. Band—the thickness or width of several rovings, yarns, or tows as it is applied to a mandrel or tool. A common term used in filament winding.
9. Tape—a composite product form in which a large number of parallel filaments (e.g., tows) are held together with an organic matrix material (e.g., epoxy). The length of the tape, in the direction of the fibers, is much greater than the width, and the width is much greater than the thickness. Typical tape product forms are several hundred feet long, 6–60 inches wide, and 0.005–0.010 inch thick.
10. Woven cloth—another composite product form made by weaving yarns or tows in various patterns to provide reinforcement in two directions, usually 0 degree and 90 degree. Typical two-dimensional (2D) woven cloth is 700 feet long, 24–60 inches wide, and 0.010–0.015 inch thick.

Some examples of fiber forms are shown in Fig. 2.11. Those are by no means comprehensive but offer examples of most commonly used forms in the aerospace industry. Now that we have some basic terminology defined, we introduce some of the commonly used fiber materials.

5.2 Fiber materials

5.2.1 Fiberglass

Fiberglass is often used for secondary structures on aircraft, such as fairings, radomes, and wing tips. Fiberglass is also used for helicopter rotor blades.

There are several types of fiberglass used in the aviation industry. Electrical glass (known as E-glass) is identified as such for electrical applications. It has high resistance to current flow. E-glass is made from borosilicate glass. S-glass and S2-glass identify as structural fiberglass that have a higher strength than E-glass. S-glass is produced from magnesia-alumina-silicate. Advantages of fiberglass are lower cost than other composite materials, chemical, or galvanic corrosion resistance, and electrical properties (fiberglass does not conduct electricity). Fiberglass has a white color and is available as a dry fiber fabric or prepreg material form.

5.2.2 Kevlar

Kevlar is DuPonts name for aramid fibers. Aramid fibers are light weight, strong, and tough. Two types of aramid fiber are used in the aviation industry. Kevlar 49 which has a high stiffness and Kevlar 29 which has a low stiffness. An advantage of aramid fibers is their high resistance to impact damage, so they are often used in areas prone to impact. The main disadvantage of aramid fibers is their general weakness in compression and hygroscopy. Service reports have indicated that some parts made from Kevlar absorb up to 8% of their weight in water. Therefore, parts made from aramid fibers need to be protected from the environment. Another disadvantage is that Kevlar is difficult to drill and cut. The fibers fuzz easily, and special scissors are needed to cut the material. Kevlar is often used for military ballistic and body armor applications. It has a natural yellow color and is available as dry fabric and prepreg material. Bundles of aramid fibers are not sized by the number of fibers like carbon or fiberglass but by weight.

5.2.3 Carbon/Graphite

One of the first distinctions to be made among fibers is the difference between carbon and graphite fibers, although the terms are frequently used interchangeably. Carbon and graphite fibers are based on graphene (hexagonal) layer networks present in carbon. If the graphene layers, or planes, are stacked with three-dimensional order, the material is defined as graphite. Usually extended time and temperature processing is required to form this order, making graphite fibers more expensive. If bonding between planes is weak and disorder frequently occurs such that only two-dimensional ordering within the layers is present, this material is defined as carbon.

Carbon fibers are very stiff and strong, 3–10 times stiffer than glass fibers. Carbon fiber is used for structural aircraft applications, such as floor beams, stabilizers, flight controls, and primary fuselage, and wing structures. Advantages include its high strength and corrosion resistance. Disadvantages include lower conductivity than aluminum; therefore, a lightning protection mesh or coating is necessary for aircraft parts that are prone to lightning strikes. Another disadvantage of carbon fiber is its high cost. Carbon fiber is gray or black in color and is available as dry fabric and prepreg material. Carbon fibers have a high potential for causing galvanic corrosion when used with metallic fasteners and structures.

5.2.4 Boron

Boron fibers are very stiff and have a high tensile and compressive strength. The fibers have a relatively large diameter and do not flex well; therefore, they are available only as a prepreg tape product. An epoxy matrix is often used with the boron fiber. Boron fibers are used to repair cracked aluminum aircraft skins, because the thermal expansion of boron is close to aluminum and there is no galvanic corrosion potential. The boron fiber is difficult to use if the parent material surface has a contoured shape. The boron fibers are very expensive and can be hazardous for personnel. Boron fibers are used primarily in military aviation applications.

5.2.5 Ceramic

Ceramic fibers are used for high-temperature applications, such as turbine blades in gas turbine engines. The ceramic fibers can be used to temperatures up to 2200 °F.

5.2.6 Lightning protection

An aluminum airplane is quite conductive and can dissipate the high currents resulting from a lightning strike. Carbon fibers are 1000 times more resistive than aluminum to current flow, and epoxy resin is 1,000,000 times more resistive (i.e., perpendicular to the skin). The surface of an external composite component often consists of a ply or layer of conductive material for lightning strike protection because composite materials are less conductive than aluminum. Many different types of conductive materials are used ranging from nickel-coated graphite cloth to metal meshes to aluminized fiberglass to conductive paints. The materials are available for wet layup and as prepreg.

When damage occurs to an external part that had a lighting protection material, in addition to a normal structural repair, the technician must also recreate the electrical conductivity designed into the part. These types of repairs generally require a conductivity test to be performed with an ohm meter to verify minimum electrical resistance across the structure. When repairing these types of structures, it is extremely important to use only the approved materials from authorized vendors, including items such as potting compounds, sealants, adhesives, and so forth.

6 Matrix materials

The role of the matrix is to bind the fibers together in an orderly manner and protect them from the environment. The matrix transfers loads to the fibers, which is critical in compression loading by preventing premature failure due to fiber micro-buckling. The matrix also provides the composite with toughness, damage tolerance, and impact and abrasion resistance. The properties of the matrix also determine the maximum usage temperature, resistance to moisture and fluids, and thermal and oxidative stability.

Polymeric matrices for advanced composites are classified as either thermosets or thermoplastics. Thermosets are low-molecular-weight, low-viscosity monomers (~2,000 cP) that are converted during curing into three-dimensional cross-linked structures that are infusible and insoluble. Cross-linking results from chemical reactions that are driven by heat generated either by the chemical reactions themselves (i.e., exothermic heat of reaction) or by externally supplied heat. As curing progresses, the reactions accelerate and the available volume within the molecular arrangement decreases, resulting in less mobility of the molecules and an increase in viscosity. After the resin gels and forms a rubbery solid, it cannot be remelted. Further heating causes additional cross-linking until the resin is fully cured. Since cure is a thermally driven event requiring chemical reactions, thermosets are characterized as having rather long processing times. In contrast, thermoplastics are not chemically cross-linked with heat and, therefore, do not require long-cure cycles. Thermoplastics are high molecular-weight polymers that can be melted, consolidated and then cooled. They may also be subsequently reheated for forming or joining operations; however, due to their inherently high-viscosity and high-melting points, high temperatures and pressures are normally required for processing.

Thermoset resins and autoclave curing have been the foundation of aerospace composites since their introduction to aircraft. During the 1960s and 1970s, most aero composite manufacturers borrowed heavily from the wet layup techniques used in the boat-building industry. But the inconsistencies and variability of the wet layup process gave way in the 1980s to more consistent, repeatable hand layup of prepreg materials. In the mid 1990s, prepreg layup gave way to more productive automated tape laying and fiber placement technologies. In all cases, the autoclave was considered necessary to ensure that laminates met void content targets. The first decade of this century, however, saw out-of-autoclave (OOA) processing techniques attract interest, with promises of speedier production and lower fabrication costs.

As a consequence of this paradigm shift toward process/cost efficiency, reinforced thermoplastics now appear on the verge of capturing a significant piece of the aerospace market on the strength of a significant distinction. Unlike thermosets, thermoplastics do not need to crosslink (cure). These polymers shape easily under sufficient heat and simply harden and maintain those shapes when cooled. Furthermore, they retain their plasticity and this characteristic offers intriguing possibilities for both faster and more innovative composite processing techniques compared to their thermoset counterparts.

The introduction of OOA processes and thermoplastics to the aerospace industry has complicated the aero manufacturer's options of material/process options. OOA processing can involve either thermosets or thermoplastics. At the same time, manufacturing with thermoplastic composites (TPCs) can sometimes involve the use of an autoclave. Now let us introduce more specific materials used for each type of polymer material.

6.1 Thermosetting resins

Resin is a generic term used to designate the polymer. The resin, its chemical composition, and physical properties fundamentally affect the processing, fabrication, and ultimate properties of a composite material. Thermosetting resins are the most diverse and widely used of all man-made materials. They are easily poured or formed into any shape, are compatible with most other materials, and cure readily (by heat or catalyst) into an insoluble solid. Thermosetting resins are also excellent adhesives and bonding agents.

6.1.1 Curing stages of resins

Thermosetting resins use a chemical reaction to cure. There are three curing stages, which are called A, B, and C.

A stage: The components of the resin (base material and hardener) have been mixed but the chemical reaction has not started. The resin is in the A stage during a wet layup procedure.

B stage: The components of the resin have been mixed and the chemical reaction has started. The material has thickened and is tacky. The resins of prepreg materials are in the B stage. To prevent further curing the resin is placed in a freezer at 0 °F. In the frozen state, the resin of the prepreg material stays in the B stage. The curing starts when the material is removed from the freezer and warmed again.

C stage: The resin is fully cured. Some resins cure at room temperature and others need an elevated temperature cure cycle to fully cure.

There are many different types of thermoset resins used in the aerospace industry that are summarized below:

6.1.2 Polyester resins

Polyester resins are relatively inexpensive, fast processing resins used generally for low-cost applications. They are used in interior parts of the aircraft given their low-smoke producing ability. Fiber-reinforced polyesters can be processed by many methods. Common processing methods include matched metal molding, wet layup, press (vacuum bag) molding, injection molding, filament winding, pultrusion, and autoclave processing.

6.1.3 Vinyl ester resin

The appearance, handling properties, and curing characteristics of vinyl ester resins are the same as those of conventional polyester resins. However, the corrosion resistance and mechanical properties of vinyl ester composites are much improved over standard polyester resin composites.

6.1.4 Phenolic resin

Phenol-formaldehyde resins were first produced commercially in the early 1900s for use in the commercial market. Urea formaldehyde and melamine-formaldehyde appeared in the 1920–1930 as a less expensive alternative for lower temperature use. Phenolic resins are used for interior components because of their low smoke and flammability characteristics.

6.1.5 Epoxy

Epoxies are polymerizable thermosetting resins and are available in a variety of viscosities from liquid to solid. There are many different types of epoxies, and the engineer should use the product specification to select the correct type for a specific application. Epoxies are used widely in resins for prepreg materials and structural adhesives. The advantages of epoxies are high strength and modulus, low levels of volatiles, excellent adhesion, low shrinkage, good chemical resistance, and ease of processing. Their major disadvantages are brittleness and the reduction of properties in the presence of moisture. The processing or curing of epoxies is slower than polyester resins. Processing techniques include autoclave molding, filament winding, press molding, vacuum bag molding, resin transfer molding, and pultrusion. Curing temperatures vary from room temperature to approximately 350 °F. The most common cure temperatures range between 250 and 350 °F

6.1.6 Polyimides

Polyimide resins excel in high-temperature environments where their thermal resistance, oxidative stability, low coefficient of thermal expansion, and solvent resistance benefit the design. Their primary uses are circuit boards and hot engine and airframe structures. A polyimide may be either a thermoset resin or a thermoplastic. Polyimides require high-cure temperatures, usually in excess of 550 °F. Consequently, normal epoxy composite bagging materials are not suitable, and steel tooling becomes a necessity. Polyimide bagging and release films, such as Kapton® are used. It is extremely important that Upilex® replace the lower cost nylon bagging and polytetrafluoroethylene (PTFE) release films common to epoxy composite processing. Fiberglass fabrics must be used for bleeder and breather materials instead of polyester mat materials due to the low-melting point of polyester.

6.1.7 Polybenzimidazoles (PBI)

Polybenzimidazole resin is extremely high-temperature resistant and is used for high-temperature materials. These resins are available as adhesive and prepreg.

6.1.8 Bismaleimides (BMI)

Bismaleimide resins have a higher temperature capability and higher toughness than epoxy resins, and they provide excellent performance at ambient and elevated temperatures. The processing of bismaleimide resins is similar to that for epoxy resins. BMIs are used for aero engines and high-temperature components.

BMIs are suitable for standard autoclave processing, injection molding, resin transfer molding, and SMC among others.

6.2 Thermoplastic resins

Thermoplastic materials can be softened repeatedly by an increase of temperature and hardened by a decrease in temperature. Processing speed is the primary advantage of thermoplastic materials. Chemical curing of the material does not take place during processing, and the material can be shaped by molding or extrusion when it is soft.

6.2.1 Semicrystalline thermoplastics

Semi-crystalline thermoplastics possess properties of inherent flame resistance, superior toughness, good mechanical properties at elevated temperatures and after impact, and low-moisture absorption. They are used in secondary and primary aircraft structures. Combined with reinforcing fibers, they are available in injection molding compounds, compression-moldable random sheets, unidirectional tapes, prepregs fabricated from tow, and woven prepregs. Fibers impregnated in semi-crystalline thermoplastics include carbon, nickel-coated carbon, aramid, glass, quartz, and others.

6.2.2 Amorphous thermoplastics

Amorphous thermoplastics are available in several physical forms, including films, filaments, and powders. Combined with reinforcing fibers, they are also available in injection molding compounds, compressive moldable random sheets, unidirectional tapes, and woven prepregs. The fibers used are primarily carbon, aramid, and glass. The specific advantages of amorphous thermoplastics depend upon the polymer. Typically, the resins are noted for their processing ease and speed, high-temperature capability, good mechanical properties, excellent toughness and impact strength, and chemical stability. The stability results in unlimited shelf life, eliminating the cold storage requirements of thermoset prepregs.

There is a wide range of thermoplastic materials now used in advanced composites components for the aerospace industry. Six general classes of thermoplastics are seen most frequently:

6.2.3 Polycarbonates (PC)

Polycarbonate is a dimensionally stable, transparent thermoplastic with a structure that allows for outstanding impact resistance. With high-performance properties, polycarbonate is the leading plastic material for various applications that demand high-functioning temperatures and safety features. Popular uses of polycarbonate can include aircraft parts, data storage devices, dome lights, eye protection, multiwall sheets, electronic components and more. An example use is shown in Fig. 2.12.

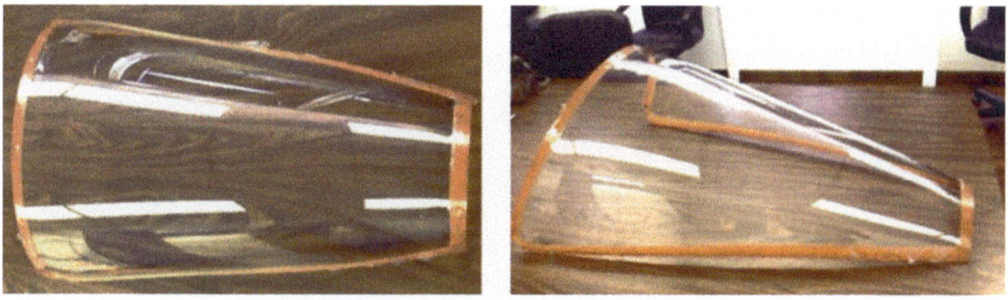

FIGURE 2.12 **Example use of PC material in aircraft canopies.**

6.2.4 Polyamides (nylon, PA-6, PA-12)

Polyamide nylon is a semi-crystalline thermoplastic with low-density and high-thermal stability. Polyamides are among the most important and useful technical thermoplastics due to their outstanding wear resistance, good coefficient of friction, and very good temperature and impact properties. In addition, nylon polyamide exhibits very good chemical resistance and is an especially oil-resistant plastic. This excellent balance of properties makes the PA polymer an ideal material for metal replacement in applications, such as automotive parts, industrial valves, railway tie insulators and other industry uses, whose design requirements include high strength, toughness, and weight reduction. Nylon plastic shows a propensity to absorb moisture and thus have poorer dimensional stability than other engineering plastics.

6.2.5 Polyphenylene sulfide (PPS)

Polyphenylene sulfide (PPS) is a high-performance, engineering thermoplastic characterized by an unusual combination of properties. These properties range from high-temperature performance to dimensional stability and excellent electrical insulation properties. An example use of this material is on the tail of the Gulfstream 650 as shown in Fig. 2.13.

6.2.6 Polyetherimide (PEI)

Polyetherimide (PEI) is an amorphous engineering thermoplastic known to exhibit high-temperature resistance, outstanding mechanical, and electrical properties. Polyimides are a relatively new class of specialty plastic materials that are characterized by:

- High strength-to-weight ratio
- Thermo-oxidative stability
- Excellent mechanical properties
- High temperatures resistance and more...

Polyetherimides have been developed to overcome challenges associated with polyimides, that is, this polymer family is not readily melt processable,

FIGURE 2.13 **The Gulfstream G650's tail rudder and elevators are made of carbon fiber/polyphenylene sulfide (PPS) thermoplastic composite [20].**

FIGURE 2.14 **Aircraft part made from Ultem material [21].**

and finished parts tend to be rather expensive. Polyetherimide was first developed in 1982 by General Electric Company (now known as SABIC) under the trade name ULTEM resin. An example part made from this material is shown in Fig. 2.14.

6.2.7 Polyether ether ketone (PEEK)

Polyether ether ketone, better known as PEEK, is a high-temperature thermoplastic. This aromatic ketone material offers outstanding thermal and combustion characteristics and resistance to a wide range of solvents and proprietary fluids. PEEK can also be reinforced with glass and carbon.

6.2.8 *Polyetherketoneketone (PEKK)*

PEKK has an extremely high-melting point (580–680 °F depending on the grade) and provides excellent resistance to chemicals and abrasion. Reinforced with carbon fibers, it is as rigid as some metals, but significantly lighter. It is non-flammable and does not generate toxic fumes. All this plus it is easily shaped above its melting point. More conventionally, it can also be injected into molds or extruded to produce tubes or films.

7 Thermal analysis

Thermal analysis is the general term given to a group of analytical techniques that measure the properties of a material as it is heated or cooled. Techniques such as differential scanning calorimetery (DSC), thermal mechanical analysis (TMA) and thermogravimetric analysis (TGA) are used to determine the degree of cure, the rates of cure, heat levels in the reactions, melting points of thermoplastics and thermal stability.

7.1 Glass transition temperature

The cured glass transition temperature (Tg) of a polymeric material is the temperature at which it changes from a rigid glassy solid into a softer, semiflexible material. At this point the polymer structure is still intact but the cross-links are no longer locked in position. The Tg determines the maximum use-temperature for a composite or an adhesive, above which the material exhibits significantly reduced mechanical properties. Since most thermoset polymers absorb moisture that severely depresses the Tg, the actual use temperature should be about 50 °F lower than the wet or saturated Tg. The cured glass transition temperature can be determined by several methods, such as TMA, DSC, or Dynamic Mechanical Analysis (DMA), all of which give slightly different results because each measures a different property of the resin. Schematic outputs from these three methods are shown in Fig. 2.15.

7.2 Material characterization

The following section presents the methodical approach developed by researchers in order to investigate the processing properties of thermoset resins which will be discussed in further detail later on in this book. The process starts by performing a set of thermal stability tests to determine the resin temperature-processing window. This is followed by investigating the degree of cure as a function of time and temperature to determine the appropriate cure profile for the specific resin. Both the glass transition temperature and viscosity are then

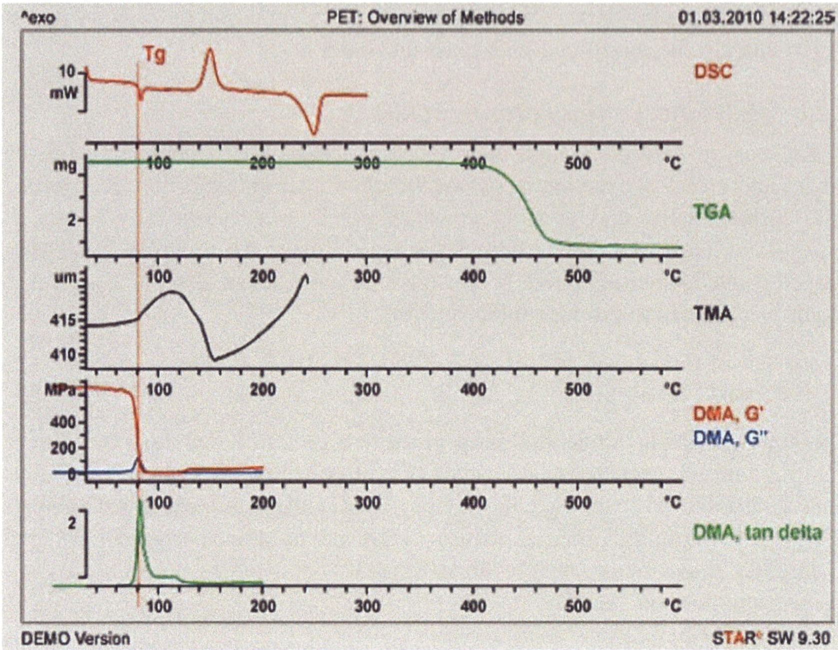

FIGURE 2.15 **Comparison between three different thermal analysis to compute Tg [22].**

measured as a function of the degree-of-cure. With the resin rheological behavior available, the cure shrinkage is determined as a function of the degree-of-cure. We can also find the coefficient of thermal expansion and the resin modulus as a function of the glass transition temperature. The process concludes by developing a material constitutive model that can be implemented in any finite element software to predict the final properties of a composite structure as a function of the cure cycle used. A summary of this process is shown in Fig. 2.16.

Details of each step are summarized as follows.

7.2.1 Step 1: thermal stability

Thermal stability tests are carried out on a TGA apparatus similar to that shown in Fig. 2.17. The process is focused on determining the weight loss of a resin sample going through a typical cure cycle to what it will observe during fabrication of a part. A temperature ramp at 20 °F/min from 75–1300 °F should be applied to a 12.24 mg sample. The sample is put under nitrogen from 75 to 1022 °F and air from 1022 to 1300 °F. A 3-hour isothermal test can also be performed at 360 °F under nitrogen. Weight loss will need to be measured after this cycle. If a weight loss of less than 2.1% is realized after a 3 hours isothermal experiment at 360 °F then that confirms the thermal stability of the sample. Therefore, the resin is not subjected to significant degradation during its typical cure cycle

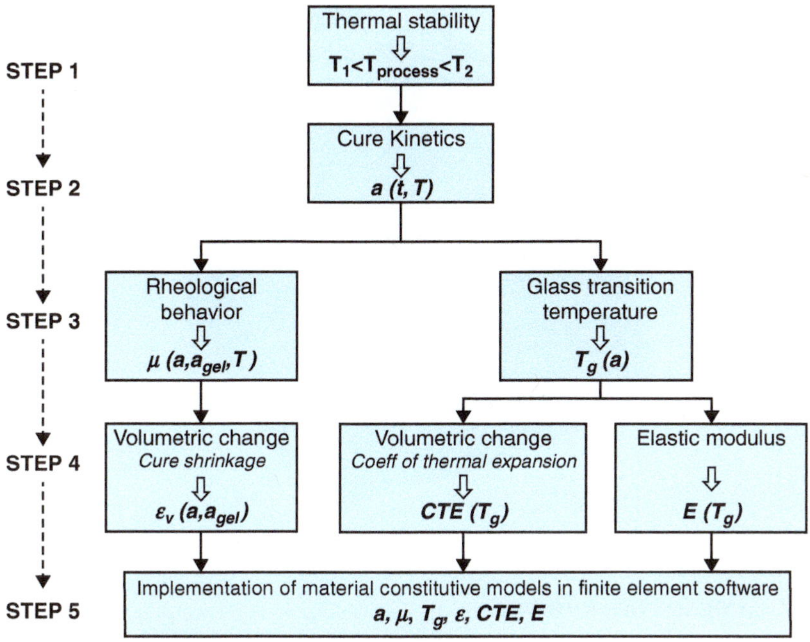

FIGURE 2.16 **Characterization of thermoset resins.**

FIGURE 2.17 **Thermal gravimetric analyzer used during thermal stability analysis [23].**

FIGURE 2.18 **Modulated differential scanning calorimeter [24].**

7.2.2 Step 2: cure kinetics

The resin cure kinetics is measured with a modulated differential scanning calorimeter similar to that shown in Fig. 2.18. The primary goal of this process is to understand the cure rate as a function of the degree-of-cure. That will be accomplished by performing both isothermal scans and dynamic scans. The dynamic scans measure the total heat of reaction released during the cure, whereas isothermal scans are used to monitor the heat flow during a series of isothermal cures. The measured heat, generated while the resin reacts, can be related to its cure rate and the degree-of-cure. Dynamic scans with heating rates of 1 °F/min and 2 °F/min, from 75 to 480 °F, and isothermal scans at 320 °F, 338 °F, 360 °F, and 375 °F, can be performed on uncured neat resin sample. Isothermal tests are followed by a dynamic ramp in temperature in order to measure the residual heat of reaction. With this information at hand, the cure kinetic model is then the relation that expresses the cure rate as a function of the degree-of-cure. A typical heat flow from a dynamic scanning test is shown in Fig. 2.19.

7.2.3 Step 3a: rheological behavior

Rheological measurements are performed using a rheometer such as that shown in Fig. 2.20. The objective of this step is to determine the gel point and minimum viscosity using different dynamic and isothermal cycles. First step involves the understanding of the linear viscoelastic region for the specific neat resin one is dealing with. A strain sweep is performed followed by a time sweep test to determine the boundaries related to the resin system used. That is followed by dynamic scans at heating rates of 1 °F/min, 2 °F/min, and 3 °F/min and isothermal scans at temperatures of 175 °F, 340 °F, 360 °F, and 375 °F on uncured neat

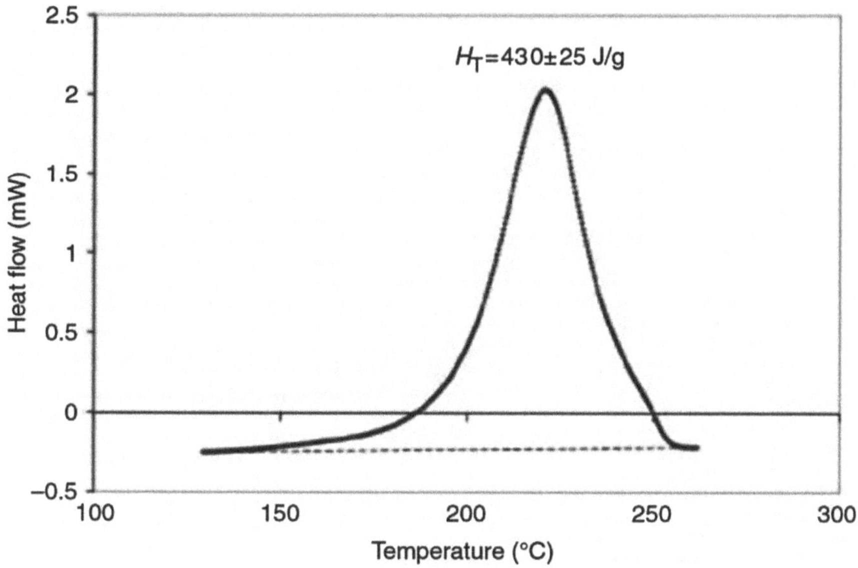

FIGURE 2.19 **Typical heat flow of a dynamic scanning calorimetry dynamic test at 2 °C/ min [25].**

FIGURE 2.20 **Example of a rheometer used for this analysis [26].**

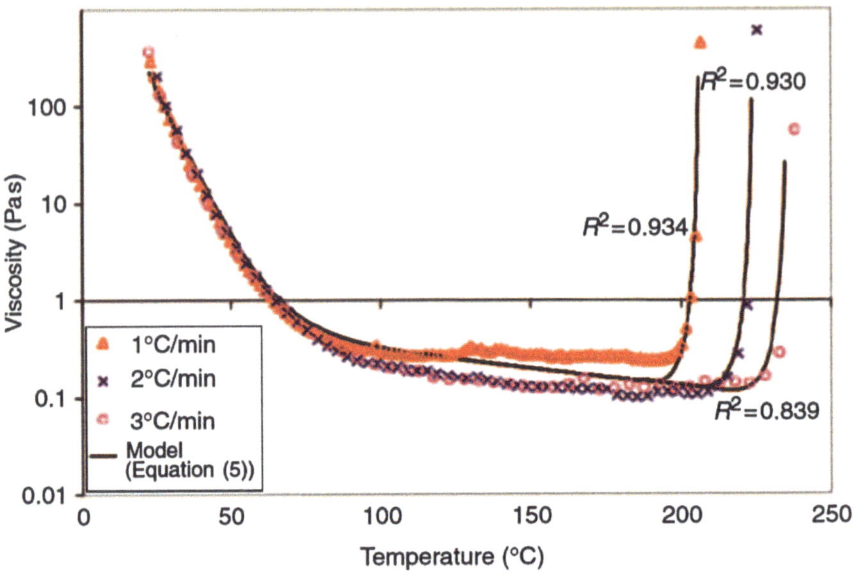

FIGURE 2.21 **Evolution of the measured and predicted viscosity with temperature for rheological dynamic tests at three temperature rates [25].**

resin sample, using a 40 mm parallel plate geometry in oscillatory mode within the linear viscoelastic region of the resin (e.g., 15% strain) and 1Hz. A summary of a typical viscosity behavior for thermoset resins as a function of temperature and at different heating rates is shown in Fig. 2.21. It can be observed that the resin viscosity decreases as the temperature increases, until it reaches a minimum value. After a certain time, the viscosity increases quickly. This sharp increase in viscosity corresponds to the gel transition. The equality between the storage and loss shear moduli, G' and G" is used as criterion to determine the gel point. It can also be seen that the curing temperature of 360 °F, the gel point occurred around 72 minutes on average.

7.2.4 Step 3b: glass transition temperature

There is a myriad of ways to compute the Tg as described earlier (e.g., MDSC, TMA, and the rheometer in torsion mode). The analyst can use any of these techniques for the evaluation. With the DSC technique, Tg is identified by a step change in the specific heat, during the dynamic ramp following an isothermal test, whereas with the TMA, it is identified by a change in CTE.

7.2.5 Step 4: cure shrinkage

The cure shrinkage can be found using the modified rheology method, which is a simple test, easy to setup. It measures the shrinkage after the gel point using a rheometer with parallel plate geometry as shown in Fig. 2.22. A controlled

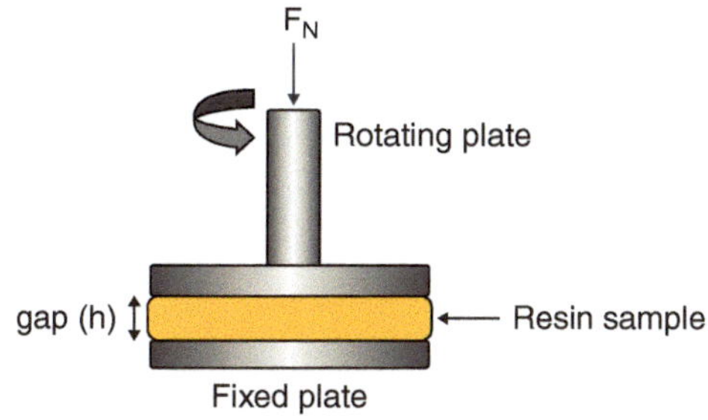

FIGURE 2.22 **Modified rheology method used to compute cure shrinkage.**

normal force is applied to maintain the contact between the plates and the resin sample while the gap variation between the parallel plates is measured. The linear shrinkage is first determined based on the change in gap between the parallel plates.

7.2.6 Step 4b: coefficient of thermal expansion

Tests are carried out using a TMA as shown in Fig. 2.23. Both fully and partially cured resin samples should be used in this evaluation to understand the impact of degree of cure on CTE. The samples can be prepared with the rheometer with 40mm plates. The curing temperatures considered can be 320 °F, 340 °F, and 360 °F, and the influence of the degree-of-cure is observed after the gel point, from $\alpha = 0.8$ to $\alpha = 1$. Then the 40 mm disk is cut in smaller samples of about 5 mm by 5 mm and 1mm thick. Three to four cycles from room temperature

FIGURE 2.23 **Thermomechanical analyzer [27].**

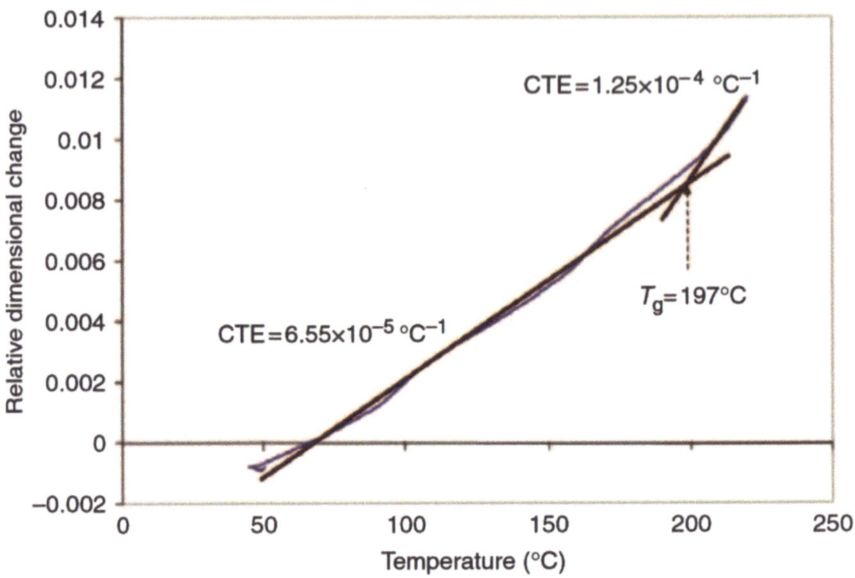

FIGURE 2.24 **Resin relative dimensional change during the heating part of cycle 3 at 3 °C/min [25].**

up to 482 °F, then back down to room temperature, at a rate of 3 °F/min are performed for each sample. A normal force of 0.05N is applied on the probe in order to maintain contact with the sample. Fig. 2.24 show a typical experiment applied on one sample and the determination of its CTE.

7.2.7 Step 4c: elastic modulus

The torsion mode of the rheometer is used to capture the evolution of the elastic modulus with the cure. The shear moduli can be related to the degree-of-cure as well. As shown in Fig. 2.25, the evolution of the modulus is very sensitive the glass transition temperature. A significant decrease in the modulus is observed as soon as the sample reached the glass transition region and changed from glassy to rubbery state (T > Tg). Then the modulus remained low until vitrification.

7.2.8 Step 5: constitutive relation

Once the data is available, one can create a constitutive relation that feeds into a user subroutine (e.g., UMAT) to express the stresses and strain generated in a model with a more accurate prediction considering the impact of the cure rate, glass transition temperature, CTE, including the resin modulus which can also be expressed as a model developed as a function of the difference between the instantaneous cure and glass transition temperature. Later in Chapter 7 we show how this can be implemented using the commercial software Abaqus in combination

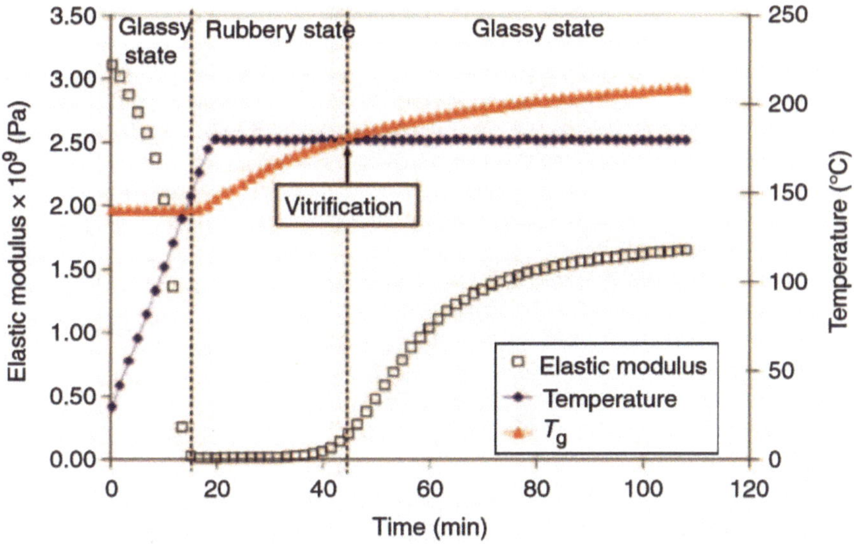

FIGURE 2.25 Evolution of the elastic modulus with temperature and time for a resin sample with an initial degree-of-cure of 85% [25].

with Compro which is a plug-in that was created with different composite materials characterized that can be leveraged for any analysis of this type.

Chapter questions

1. What is thermal analysis and how is it used?
2. Name three methods used to find Tg experimentally?
3. Name five different types of fibers used in the aerospace industry?
4. Name the two primary types of matrix systems used and the main distinction between both?
5. Define degree of cure in the context of composite materials?
6. Which fiber form is used primarily in impact application?

References

[1] https://www.rockwestcomposites.com/800881.
[2] https://www.carbon.ee/en/p/blog.
[3] http://canyouactually.com/a-giant-braiding-machine-weaving-together-fiberglass-is-hypnotizing-to-watch/
[4] https://www.defender.com/product.jsp?id=17388
[5] https://bindatex.com/chopped-composite-fibre/
[6] Balasubramanian M. Composite materials and processing, CRC Press, Boca Raton 2017, https://www.crcpress.com/Composite-Materials-and-Processing/Balasubramanian/p/book/9781138076877.

[7] Composite materials: fabrication handbook #1 (Composite Garage Series) John Wanberg, 2009, https://www.amazon.com/Composite-Materials-Fabrication-Handbook-Garage/dp/1929133766.
[8] Chawla KK., Composite materials: science and engineering (Materials Research and Engineering). 3rd ed., https://www.springer.com/gp/book/9781493950157#otherversion=9780387743646.
[9] Daniel IM., Ishai O., Engineering mechanics of composite materials, 2nd ed. https://www.amazon.com/Mechanics-Composite-Materials-Science-Engineering/dp/156032712X.
[10] A. Baker A., Dutton S., Kelly D. Composite materials for aircraft structures, 2nd ed. (AIAA Education Series), https://www.amazon.com/Composite-Materials-Aircraft-Structures-Education/dp/1563475405.
[11] Barbero EJ. Introduction to composite materials design, 3rd ed., https://www.amazon.com/Introduction-Composite-Materials-Design-Barbero/dp/1138196800.
[12] https://www.sdcomposites.com/Facilities/manufacturing.html.
[13] http://www.totalboat.com/product/vacuum-bagging-film/
[14] https://pentaxia.com/high-performance-composites/
[15] http://www.jeccomposites.com/knowledge/international-composites-news/composite-aileron-structure-cured-one-step
[16] http://zoltek.com/products/px35/prepreg/.
[17] https://www.youtube.com/watch?v=JqEd3FLZI2o.
[18] Huang HS. Influence of phase shift on the responses of woven laminated composites. Compos Struct 2015;130:143–154.
[19] https://lunainc.com/rain-repellant-coatings/canopy_coated/
[20] https://insights.globalspec.com/article/12596/thermoplastic-composites-for-aerospace-applications
[21] https://www.stratasysdirect.com/materials/thermoplastics/3d-printing-ultem
[22] https://www.azom.com/article.aspx?ArticleID=16219.
[23] https://www.msesupplies.com/products/thermal-analytical-services-tga-dta-dsc-analysis-low-cost?variant=23518973493306&utm_medium=cpc&utm_source=google&utm_campaign=Google%20Shopping&gclid=EAIaIQobChMIo6Kqvpr05AIVFtVkCh0Ujwa4EAQYASABEgIFt_D_BwE.
[24] http://www.tainstruments.com/pdf/oldDSC.pdf.
[25] http://citeseerx.ist.psu.edu/viewdoc/download?doi=10.1.1.871.7316&rep=rep1&type=pdf.
[26] http://www.tainstruments.com/pdf/brochure/2006_AR_Brochure.pdf.
[27] https://www.t-k-e.com/en/ta2900.html.

Chapter 3

Tooling materials and processing

Chapter outline

This chapter is devoted to understanding details related to the materials and processing used for the fabrication of tooling for aerospace structures. Note that the processes noted here are not explicitly for tooling only but can also be used for the purpose of building composite parts. For lamination molds (composite tools) that will go through cure cycles, the temperature variations and variables associated with heating are the primary differences when compared with assembly tooling where temperature variations do not matter as much.

We start by noting that selecting the appropriate material needs to consider many different aspects that include:

1. Cost
2. Program schedule
3. Manufacturing process
4. Lead time
5. Number of cure cycles required
6. Strength
7. Ease of repair
8. Weight
9. Heat-up rate

All those are considerations that need to be accounted for when selecting the appropriate material to use depending on the application. Characteristics of a good tool are:

1. Dimensional accuracy at both room temperature and cure temperature
2. Provides a surface that can be bagged without having any vacuum leaks

Tooling for Composite Aerospace Structures. http://dx.doi.org/10.1016/B978-0-12-819957-2.00003-1

3. Tools that are thermally stable over time through repeated cure cycles where the dimensions do not change, vacuum integrity does not degrade, and surface finish does not disintegrate
4. Durable throughout the life cycle of the tool and the quantity of parts we want to use the tool for
5. Ease of repair and modification

1 Metallic composite tooling materials

Metallic tooling provides the strongest tooling options; making them suitable for use in high mechanical loading processes such as matched-die molding. Metallic tools may be used for hundreds of production cycles with minimal regular maintenance but are time consuming and expensive to manufacture and handle. Traditional metallic-based layup tools are primarily constructed from aluminum, steel, or invar. Often, these tools are complex-welded structures requiring multiple forming, welding, machining, and heat-treatment operations to bring the tool to final geometry. Examples of such tools is shown in Fig. 3.1

Aluminum-based tooling are capable of producing high-quality surfaces and are known to be easily machined often yielding the final surface finish with only a single machining operation and minor secondary processing or finishing.

FIGURE 3.1 Metallic lamination tooling examples [1–3].

The low principal material cost of this material is another advantage it offers; however, aluminum's durability is inferior to steel and invar, which is another consideration that will need to be accounted for.

Aluminum's CTE is too high to allow for accurate high-tolerance parts to be produced at elevated cure temperatures required by many modern high-performance composite material systems. Steel is less machinable than aluminum but a fine surface finish is achievable with additional machining steps and it offers a more durable option for a longer lifecycle. Steel-based tooling can be affordable given the low cost of the raw material similar to aluminum while possessing a lower CTE but unfortunately, high tolerance aerospace applications still require tooling with even lower CTE than what steel and aluminum have to offer.

Nickel-based alloys such as invar offer the best performance of all metallic-based tooling materials in terms of durability and thermal properties. These expensive metals are the accepted standard metallic tooling material when high-tolerance and long-service life are required. The drawback in this case is that metallic tooling with low CTEs is not capable of rapid fabrication that is generally required when dealing with aerospace applications. Furthermore, the increased density of metallic tooling solutions over nonmetallic solutions lead to more weight, more thermal mass, complicated handling and high costs.

Non-metallic tooling solutions offer large performance gains when durability can be sacrificed. Applications that require short-service life that include prototype tooling, short production run tooling, and Foreign Object Damage (FOD) or other in-service damage repair tooling are common applications. Many non-metallic tooling solutions offer rapid manufacturability through a combination of simplified tooling designs and increased machinability of novel materials.

2 Nonmetallic tooling materials

Standard production tooling is required to produce large numbers of parts with minimal need for mandatory maintenance or repair. Thus, the durability of a production tooling system is paramount. Non-production tooling such as prototype tools and tooling made for repair applications are expected to yield far fewer parts than production tooling. When few part cycles are expected, less durable and more cost-efficient options can be considered. In these applications non-metallic tooling materials such as composites, ceramics, or graphite are not excluded by their lack of durability.

Graphite-based tools have been an established practice for several decades [4] and are the most widely used nonmetallic substrate material in small-scale tooling and can rapidly produce low-cost tooling since it is easily machined and has a low-raw material cost. Graphite has a well-matched CTE to carbon fibers and has been shown to produce high-tolerance composite parts at cure temperatures more than 400 °F [5].

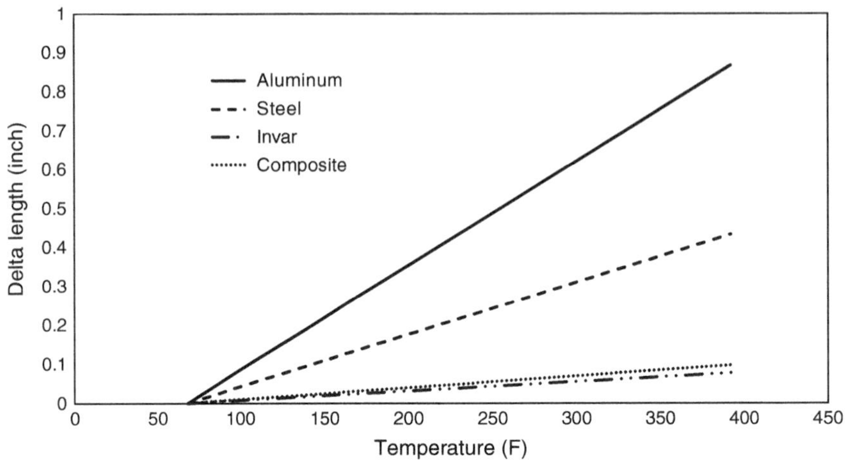

FIGURE 3.2 Length variation as a function of temperature for different materials.

The variation in length for a 16 feet-long part made from different materials is shown in Fig. 3.2. A typical cure cycle will see a temperature variation of around 360 – 70 = 290 °F which provides large change in length depending on the material selected.

The low-density and good thermal conductivity of graphite reduces thermal mass, total energy, and time required to heat the tool to the required temperature. Unfortunately, no material is perfect, and graphite has also several drawbacks to its use. Graphite tooling has poor durability and frequently requires surface repairs before subsequent use due to the low-impact strength and its tendency to fracture during curing. The dust created during machining is a possible health and safety hazard and must be contained or controlled to limit both human and machine exposure [6–8]. Graphite is a porous material, making it difficult to maintain vacuum integrity through the tool. A common solution is to apply a surface sealer directly to the tool's surface to help seal the pores and improve surface finish.

Ceramic tooling materials are more durable than graphite and can produce fine surface finishes while providing low CTE compatible with composite material systems. However, most ceramic materials are too hard for normal machining operations, and frequently are machined using abrasive type cutting tools and processes such as grinding, abrasive water-jet machining, or ultrasonic machining. Erosive processes of this nature do not achieve high material removal rates. Therefore, machining time, fabrication time, and costs are increased and lead to it being unpractical to use. Furthermore, ceramics suffer from poor thermal conductivity, which increases the time required to heat-up the tool. This can be overcome through the use of embedding heating elements within the tool but that is not a trivial process and we will touch on that later in this book. Nonetheless, this adds to the complexity and weight as well as the cost of the finished tool.

Moreover, properties of ceramics vary with density of the material and chemical composition. Ceramics with low density or special formulations that increase machinability can be machined using traditional machining processes and offer increased process economics. However, decreased density coincides with increased porosity, which can present several challenges to machining. Careful consideration of the properties the material offers and how they affect machining and final part performance is necessary.

Layup tools fabricated from composite materials made from the same or similar material to the part being cured is another option, therefore, the CTE mismatch is nearly zero which is ideal when it comes to thermal expansion issues. Composite systems used for aerospace tooling typically consist of an epoxy or bismaleimide matrix reinforced with carbon or graphite fibers. These composite systems offer low durability. Although high-quality surface finishes can be achieved from tooling made from composite materials, increased finishing (sanding/polishing), and maintenance are required to sustain the surface quality. Composite systems containing carbon or graphite fibers are abrasive in nature when machining. The high hardness and strength of the fibers along with their small diameter induce localized stress concentrations on the cutting tool edge. High tool wear rates are avoided through the use of expensive Polycrystalline Diamond (PCD) cutting tools or diamond-based coatings.

The application of where the tool is to be used weather production or prototype/R&D is a very important consideration that needs to be accounted for when making the material selection. For production, we need to make sure that it can at least build the minimum amount of parts we want to generate during the life of the program since if a new set of tools are required in the middle of the program, that will cause significant delays and high cost associated with evaluating the tool, performing the thermal surveys and so on. Another consideration is the number of design changes anticipated for a specific part or configuration. If a design change was made after a tool fabrication started then any change to the tool will be very costly as it will require an update to the tool design and if this was supposed to be fabricated at a supplier you will need to inform them to stop the tooling fabrication. If a tool was a composite it is possible to make changes by machining the surface or updating the part as shown in Fig. 3.3 to add details to support the new design.

The life dependency of nonmetallic tooling relies on several factors that include:

- **Resin type**: the resin system is used to bind all the fibers together in an orderly manner protecting them from the environment. The matrix transfers the loads to the fibers and prevents premature failure due to fiber microbuckling for example. The resin also provides the composite with toughness, damage tolerance, and impact and abrasion resistance. The combination of the resin and fiber influence the overall life cycle. When considering resin systems, BMI is considered to be more thermally stable, offers less microcracking and dimensional change but is more expensive and difficult

FIGURE 3.3 Example of tool modification to support late part design change [9].

to process. The resin properties also determine the maximum usage temperature, resistance to moisture and fluids and thermal oxidation stability.

- **Manufacturing process**: Ply consolidation during layup, debulk temperature and pressure, autoclave temperature and pressure, operating environment, and so on.
- **Machining techniques**: Shop practice, trimming plies on tool surface, demolding practices, exposure to temperatures above Tg, cleaning solvents

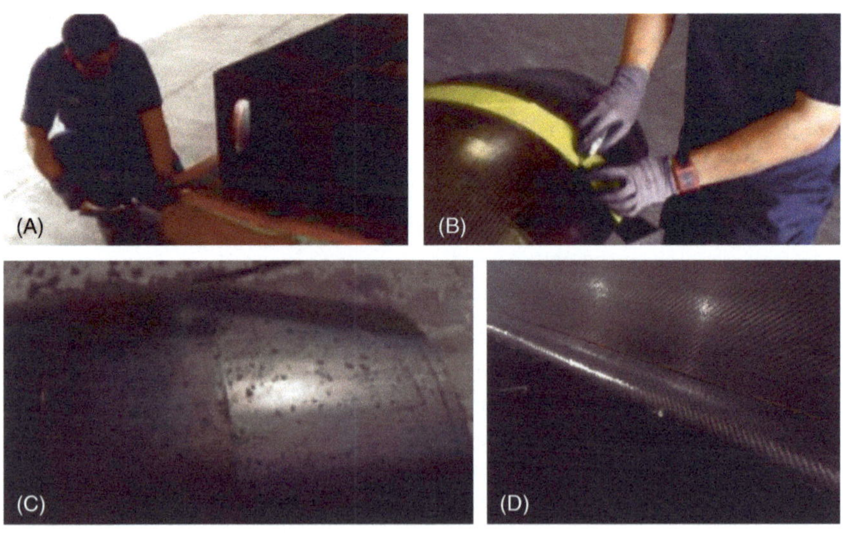

FIGURE 3.4 Examples of manufacturing techniques (A) Demolding with metal wedges. (B) Cutting on tool surface. (C) Tool exposed to moisture and UV (D) dent on composite tool.

FIGURE 3.5 Scratches from knife cuts on a lamination tool.

and practices, proper storage to avoid UV, and moisture all play a roll. An example of these techniques is shown in Fig. 3.4.

Some of the typical types of tool damage seen on nonmetallic tooling include:

- Knife cuts, scratches, nicks as shown in Fig. 3.5
- Dents from impact damage
- Delaminations due to damage, over temperature, UV exposure, or moisture exposure

It is worth noting that all composite tooling require a master mold in order to build the primary tool itself. This is certainly considered added cost when it comes to nonmetallic tooling but depending on the final tool material the mold can be made from a variety of other materials as shown in Table 3.1 and that should be another factor to consider during the planning phase of any program.

TABLE 3.1 Material options for master molds required to build composite tooling.

Tool material	Master material	Cost
CFRP infusion	Low-cost foam	Low
CFRP epoxy	High-density foam	Medium
CFRP BMI	Invar or graphite	Very high
Invar	NA	High

FIGURE 3.6 BMI tool used to fabricate the 787 fuselage sections [10].

Focusing on CFRP BMI for a moment, this combination of tools has not really been used that often in the industry. A famous example that comes to mind is the 787 tools for the fuselage section as shown in Fig. 3.6 which were made using BMI. The primary reason was the durability required to sustain thousands of parts including the ability to sustain the process of how these parts are fabricated which was Automatic Fiber Placement (AFP) so durability was extremely important in this case and a decision was made to utilize BMI for the application even though the cost was fairly high.

3 Metallic versus nonmetallic

It is clear that making a decision on which material to choose is a complex process and highly dependent on the application and situation at hand. Engineers and program managers shall consider all options prior to making a final decision on which material to use to build tools for there parts. If this decision was rushed, (and I have personally been involved in programs where this occurred) it can cause serious delays and quality issues down the road. A side-by-side comparison between CFRP and invar as material options is shown in Table 3.2. The reader is encouraged to consume the data described in this book as a reference to support the decision process for the program at hand.

4 General processing of tools

When it comes to fabrication and processing and the ability to build high-quality tools, that thought process always starts by selecting the correct material to build the tool that best fits the application but there are always general processing that takes place that should be considered and used when necessary.

TABLE 3.2 Comparison between CFRP and Invar as material options for tool fabrication.

Invar	CFRP
Very durable	Medium durability
Straightforward to repair with conventional metalworking practices	Major defects difficult to repair
Vacuum leaks can be fixed with weld	Vacuum leaks need material removal and replacement
Moderate weight	Lightest tool
Moderate thermal mass	Least thermal mass
Longer thermal cycle	Shortest cycle

One of the important processes that come to mind is related to bonding that we will discuss next. Note that all the details discussed here apply to the fabrication of composite parts as well.

4.1 Bonding

Bonding is typically used to join multiple parts together using some sort of adhesive. The adhesives come in many forms including liquids, pastes, and film adhesives as shown in Fig. 3.7. Each form has different applications where its best suited to be used.

Liquids have viscosities typically range between 100 and 6000 cps. They work best when you have a thinner bondline and provide for a higher degree of direct load transfer than pastes. The effective thickness range can be between 0.002 and 0.010 inches. Since the viscosity is low, it can run out of thicker bondlines which is certainly not desirable. Liquids tend to be more brittle and less resistant to peel loads than pastes or films. Often "liquid" adhesives are categorized as "pastes" without distinction by the various adhesive manufacturers, which is an issue and engineers need to be caution of that.

Pastes on the other hand have viscosities typically in the range > 8000 cps. They generally work better in slightly thicker bondlines with effectivity thickness range between 0.005 and 0.10 inches.

Film adhesives are similar to composite prepreg in that they need to be stored in a freezer at 0°F in order to ensure that the cure does not advance. The material comes in the form of rolls and is cut into the appropriate shapes based on the part and application. This form is considered to be the most consistent and provides the best quality bondlines.

Generally speaking, the aerospace field uses a finite set of adhesives that are shown below:

- Epoxies: wide range of high-strength adhesives available with a variety of curing and service temperatures

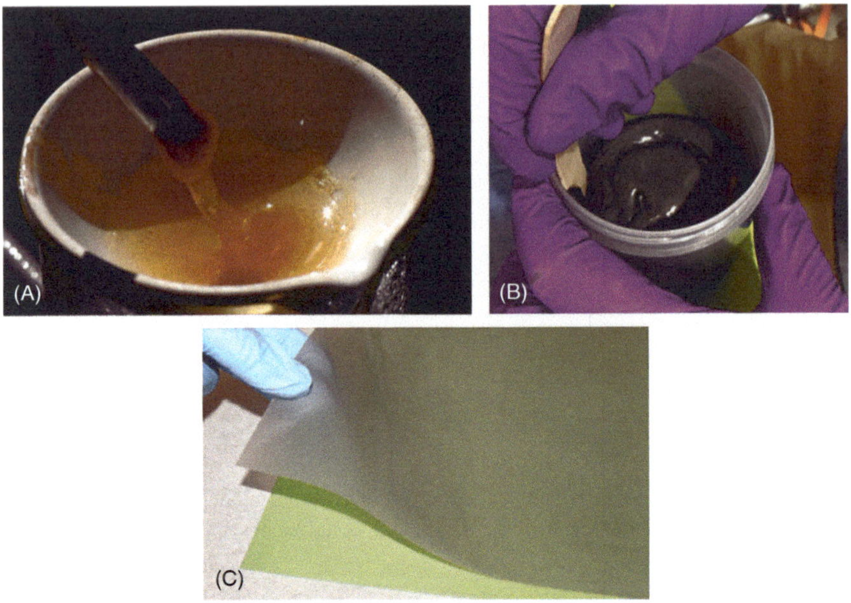

FIGURE 3.7 Different forms of adhesives used in composite material applications (A) liquid, (B) paste, and (C) film.

- Bismaleimide (BMI): high temperature cure/service (up to 600°F)
- Cyanate ester: good dielectric properties and have low CTE
- Hybrids: a combination of the several of the adhesives mentioned earlier

Some of the adhesives are used for high-performance structural prepreg material such as the epoxy film product and others are simply used to ensure that parts are attached and are not expected to carry any load. Most of the film adhesives can be stored frozen and thawed to room temperature before use and they require an elevated temperature cure cycle.

Some of the suppliers of these types of adhesives are

- 3M
- Henkel
- Master Bond
- Solvay
- Hexcel

4.2 Surface preparation for bonding

Prior to bonding any surface, it must be prepared in a way to ensure that the bonding process will be effective and no disbond will occur during operation. The primary goal is to raise the surface-free energy of the substrate to enhance wetting of the surface and to facilitate molecular cross-linking when it comes to composites in particular. In addition, the process needs to occur without damaging fibers in the laminate. The general consensus is that we need to create a surface that:

- Has high enough surface energy to form a strong interface with the adhesive
- Is mechanically strong, that is, undamaged
- Is consistent from point-to-point on the bond surface
- Is consistent from part-to-part, shift-to-shift, day-to-day

Consistency is the single greatest challenge to successful bonding and surface preparation. Predictability of bond performance is more important than ultimate properties one can argue, and engineers have to know how much strength they can count on when two or more parts are bonded. There are several methods to prepare composite surfaces that we will briefly describe next:

- Peel ply

A peel ply is a material used during the lamination process of composites and is typically located on the top most structural ply where bonding is expected to occur as shown in Fig. 3.8. The material needs to be compatible with the substrate material and several companies offer such material form with their composite prepreg.

This peel ply acts as a protective layer, which is removed immediately prior to adhesive application, which creates a clean, energized surface ready

FIGURE 3.8 Peel ply example used on composite panel.

for bonding. But note that throughout the years many issues were found using this approach [11] that were overcome by introducing other techniques such as plasma and laser.

- Abrasion

Abrasion through the available material results in a greater amount of fiber exposed at the surface. The process can be done in many different ways:

1. Hand sanding with different grit sizes 80 grit, 120 grit, 220 grit as shown in Fig. 3.9
2. Media blast via white aluminum oxide 60 grit or depending on the needs and application. Example shown in Fig. 3.10

The sanding process needs to be as minimal as possible not to remove the structural plies and should be terminated when black dust appears indicating that sanding through the fibers was achieved.

- Plasma

Given the drawback of the previous surface preparation methods, the industry several years ago endeavored to evaluate other techniques that can be utilized as an alternative. It was noted that achieving a chemical bond between substrate and adhesive, rather than mechanical interlocking, is the key to long-term bond durability [12] and achieving a repeatable, quantifiable, inspectable, and scalable surface preparation is possible via plasma treatment of the surface.

Several plasma processes exist for material surface treatment which include:

- Corona
- Vacuum plasma

FIGURE 3.9 Example of hand sanding process of composite.

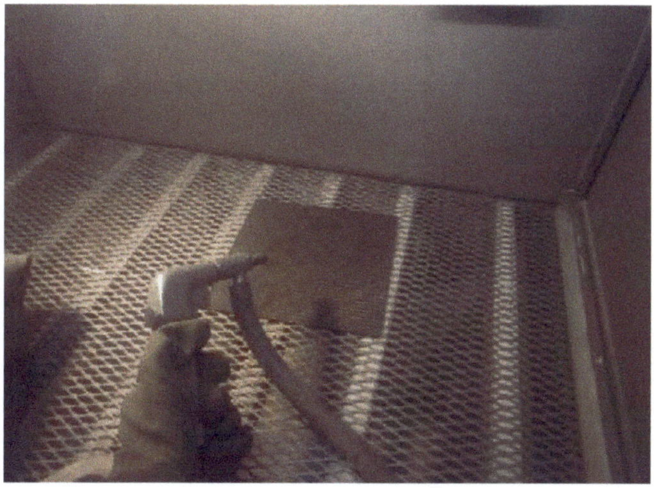

FIGURE 3.10 Example of grit blasting of composite surface.

- Flame
- Atmospheric pressure plasma
- Blown ion plasma treatment

Only atmospheric plasma process is having widespread adoption for composite surface preparation in the aerospace industry. An example of that is shown in Fig. 3.11.

The process chemically modifies the outermost surface of the composite substrate, it is able to clean the surface without damaging fibers, creates select chemical groups on the surface of the material that enhances the chemical

FIGURE 3.11 Plasma treatment of composite substrates [13].

bonding. Different gases can be used to achieve different surface polarity/chemistry.

With all these surface preparation methods and others that we have not even mentioned, how can one check surface preparation quality? One approach is by using a water break test. This test can be performed to check for sufficient cleanliness on the surface. It allows the user to check the presence of hydrophobic contaminants and residue such as oil or grease. The cleaned surface is placed in a vertical position and sprayed with drops of distilled water to wet the surface.

If the surface is clean, the water will spread out over the surface due to high-surface energy. A large contact angle is formed due to its attraction to the surface. This shows the surface is free of contamination and the surface treatment was successful. Fig. 3.12 shows an example of a good- and bad-wetting surface. There are many tools available for purchase that can be used to perform this check during production work.

4.3 Bond control and types

As stated earlier, the use of bonding can come in any form of tooling and material weather it was made from metal or composite. Another important aspect when it comes to ensuring the integrity of the bond is having a constant thickness across the bondline. For that to happen one can use a variety of different methods to control the bondline as shown below and in Fig. 3.13.

Ensuring that we have a high-quality bond will involve many different steps and one of them is having an appropriate mix of the adhesive including the correct amount of pressure and temperature, which is important for final cure of the part. A summary of the requirements is shown in Fig. 3.14.

When joining parts with adhesive it can be done using many different bond joints. Each joint has an advantage and disadvantage that are summarized in Table 3.3.

Bond joints are also loaded in many different ways as shown in Fig. 3.15. Depending on the loading conditions, the failure observed in bonds might differ and engineers need to be aware of that and design accordingly.

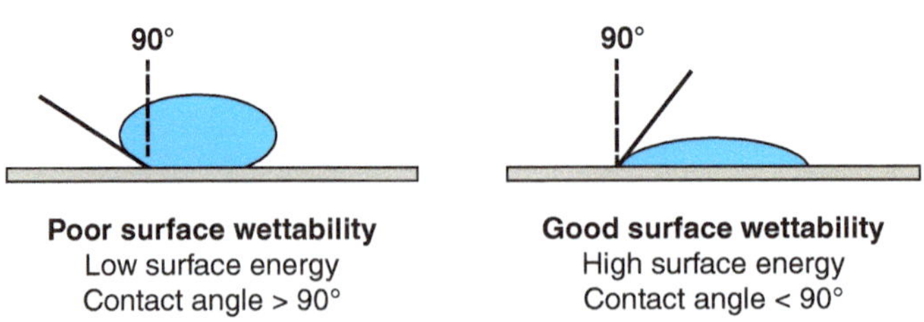

FIGURE 3.12 Water break test example [14].

FIGURE 3.13 Bondline control techniques (A) Woven scrim cloth, (B) Knit carrier, (C) Non-woven (mat), (D) glass beads.

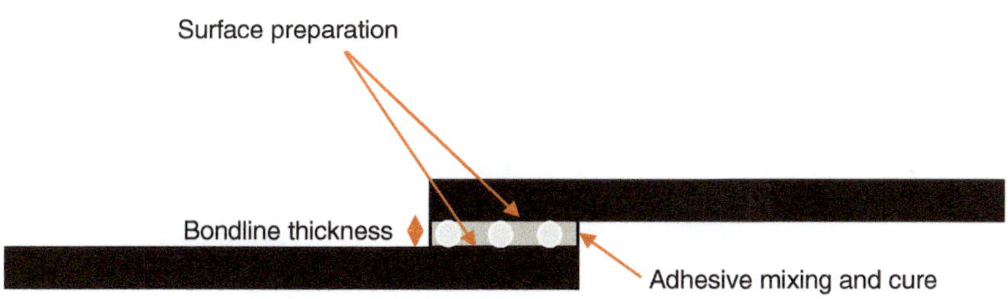

FIGURE 3.14 Requirements that need to be satisfied for a quality bond.

4.4 Failure modes

Typical failure modes in composites are summarized in Fig. 3.16. Each failure mode indicates issues with the structure. If an adhesive failure is observed after a test for example that could indicate that a surface was not prepared appropriately and is cause for concern. Details on some of these failure modes are described further.

TABLE 3.3 Different types of joint bonding.

Type joint	Advantage	Disadvantage	Thumbnail
Single lap	Good quality bond Easy to fabricate	- Delamination at termination	
Tapered single lap	Good quality bond Practical	-Difficult to mate	
Single strap lap	-Easy to fabricate	-Bond quality not as strong	
Double lap	-High quality bond strength	-Complicated to fabricate	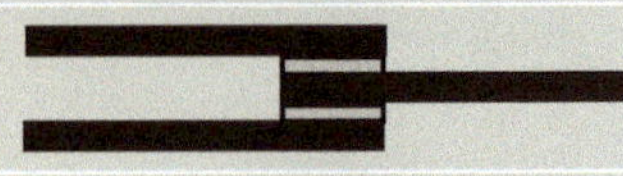
Double strap lap	-High quality bond strength	-Complicated to fabricate	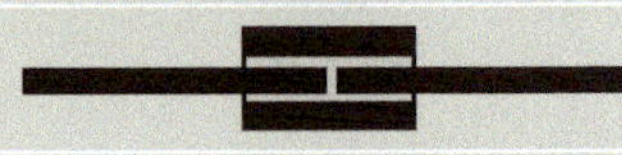
Double ta-pered strap lap	-High quality bond strength	-Complicated to fabricate	
Tapered scarf joint	-Easy to fabricate High quality bond	-Not applicable for highly loaded areas	

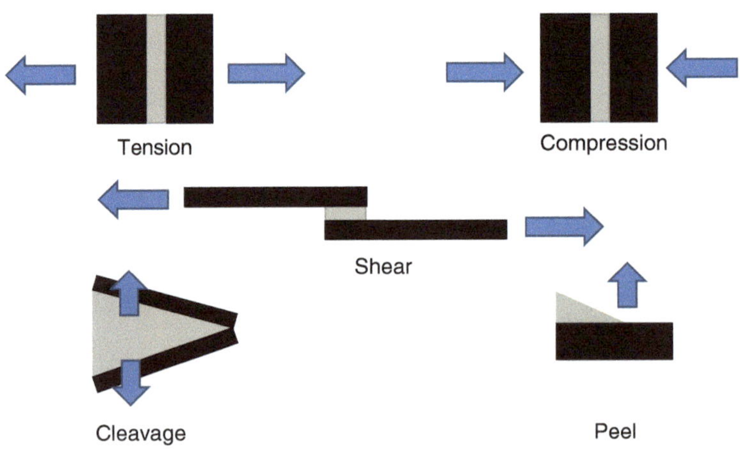

FIGURE 3.15 Bond joint loading scenarios.

Adhesive failure: Failure of a bonded joint between the adhesive and the substrate. This failure occurs primarily due to a lack of chemical bonding between the adhesive and the bonding substrate. Can be indicative of poor surface preparation, contamination or incorrect adhesive selection for the substrate material.

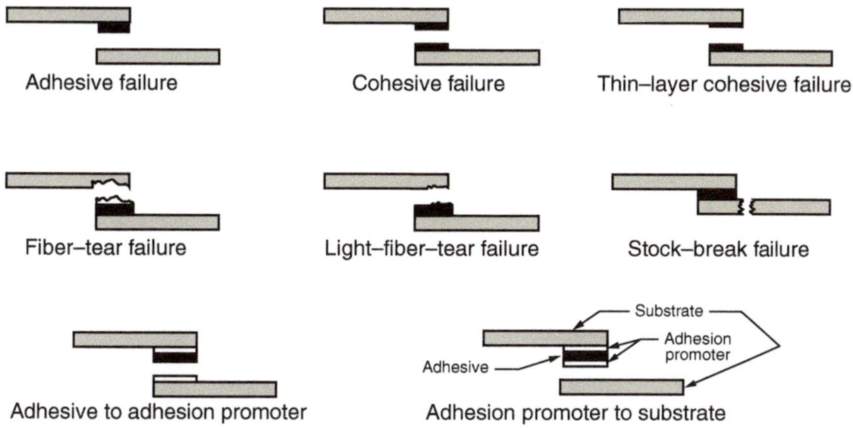

FIGURE 3.16 Composite bond failure modes.

Cohesive failure: Failure of an adhesive joint occurring primarily in the adhesive layer. Optimum type of failure in an adhesive bonded joint when failure occurs at predicted loads.

Lower failure loads are indicative of poorly cured adhesive or moisture or other contaminants present in the adhesive.

Substrate failure: Inter laminar fracture in composite structures, usually between the first and second plies adjacent to the bondline; can be common in composite laminates especially those with brittle epoxies.

All the details mentioned thus far need to be considered when selecting the adhesive type and substrate. Other factors impacting the selection of a proper adhesive are:

- Thermal conductivity of the adhesive
- Chemical compatibility
- Viscosity
- Temperature resistance
- Mechanical strength of the adhesive

In addition to all those factors, the corrosion between the substrate needs to be considered as well. A well know issue is the joints that include aluminum with composite as they are on opposite ends of the galvanic scale as shown in Fig. 3.17 and require special surface treatment before they can be bonded or assembled.

Another item worth noting is that there exist many bond assembly fixtures or jigs which are used to locate and secure mating parts for co-bond or secondary bonding operations as shown in Fig. 3.18. Those could be another variable impacting the bond quality and careful attention needs to be given to ensure successful design. The design of these fixtures may include mechanical clamping devices and/or detail locator provisions. They are designed to provide dimensionally accurate assembly at maximum process temperature.

Material	Galvanic Scale	Anodic
Clad Aluminum 7075	10	
Clad Aluminum 2024	9	
Aluminum 7075 – T6	9	
Cadmium	8	
Aluminum 2024 – T4	7	
Wrought Steel	6	
Cast Steel	6	
Tin	4	
Brass	2	
Copper	2	
Nickel	1	
300 Series 18–8 CRES	0	
Titanium	0	
Carbon Fiber	0	Cathodic

FIGURE 3.17 Galvanic corrosion scale.

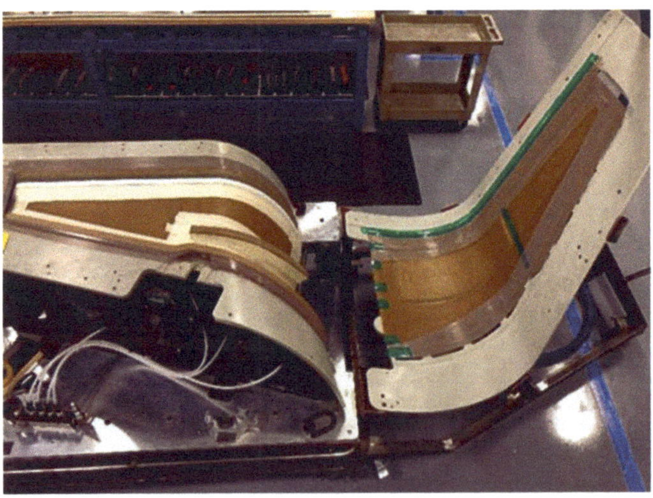

FIGURE 3.18 Bond assembly fixtures [15].

5 Release agents

The term release agent or lubricant is used to describe a wide variety of chemicals, which provide a barrier between a tool and the surface of a part being molded. There are two basic types of release agents internal and external. An internal release agent is an additive used directly into the resin formulation. The

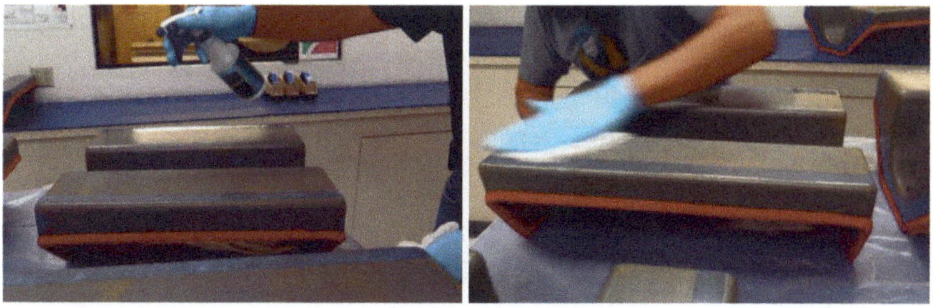

FIGURE 3.19 Tool being prepared using FreKote release agent [16].

external is applied to the exterior of the mold as shown in Fig. 3.19. Release agents are important to ensure that parts will be able to get detached and demolded easily. Some of the factors that influence the adhesion of two materials to each other are penetration, chemical reaction, surface tension, surface configuration, and polarity differences.

Release agents are an integral part of the manufacturing operations and there are many suppliers that offer such material. Since release agents can influence the part properties as well as the quality of the release, the proper material selection is crucial. The optimal material will enable the release of the part without any damage and provide many release applications and not build up on the mold.

FreKote (a Henkel Loctite brand) uses proprietary release agents that can aid in the surface release agent and work in quick production. It is important to ensure the safety when operating with this material as it can become airborne and as a result create surface contamination. This will cause subsequent steps in the operation of bonding and painting.

6 Tool material selection study

This section demonstrates a simple 2D simulation study using Raven [17] to evaluate different tool material options to fabricate a composite structure of varying cross-section shapes. For simplicity, we select a quarter sectional area as shown in Fig. 3.20 but note that the software does provide many other

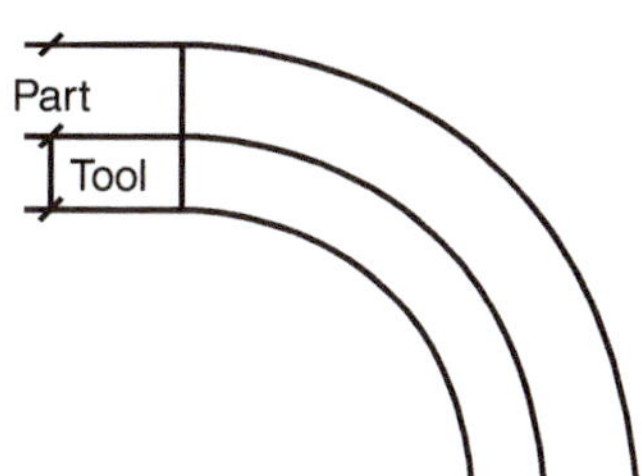

FIGURE 3.20 Cross-sectional area considered for the study.

selections to choose from. The primary objective of this study is to evaluate the heat-up rate of a tool compared to the composite and perform modifications to the tool design (e.g., thickness) to ensure uniform heating within a reasonable timeframe.

The composite material used in this study is Hexcel 8552/AS4 Tape with the composite being 0.615 inches thick and the tool selected to be 0.3 inches with a male tooling approach used. The different sides of the tool and part had specific heat transfer coefficient (HTC) applied to them as shown in Fig. 3.21 and Table 3.4.

The cure cycle applied to the part is that shown in Fig. 3.22 which mimics many of the standard cure cycles for aerospace material systems. The tooling material considered are aluminum, copper, invar 32, steel, and composite. The initial results from the simulation are shown in Table 3.5. These results show the thermal distribution at the end of the cure cycle. As shown, the lowest temperature variation is observed for the composite tool which is expected given the close match in the CTE between the part material and tool. The largest

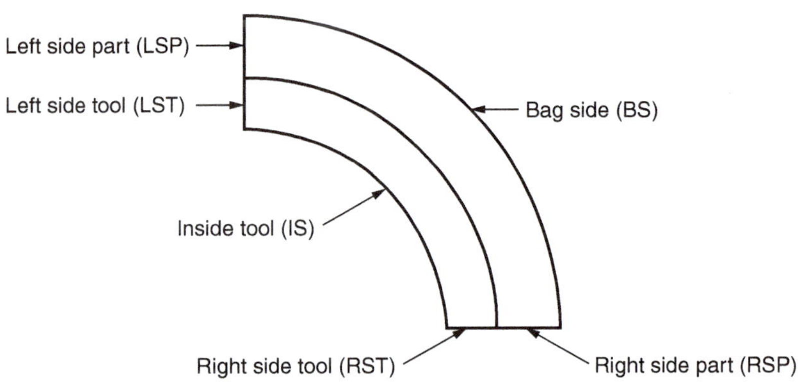

FIGURE 3.21 Different sides of the part/tool used for this study.

TABLE 3.4 Heat transfer coefficients applied to the different part/tool surfaces.

Location	Load type	Value	Units	Cycle
BS	HTC	14.08	BTU/(ft^2 hr F)	Autoclave
IS	HTC	3.52	BTU/(ft^2 hr F)	Autoclave
LSP	HTC	8.80	BTU/(ft^2 hr F)	Autoclave
LST	HTC	8.80	BTU/(ft^2 hr F)	Autoclave
RSP	HTC	8.80	BTU/(ft^2 hr F)	Autoclave
RST	HTC	8.80	BTU/(ft^2 hr F)	Autoclave

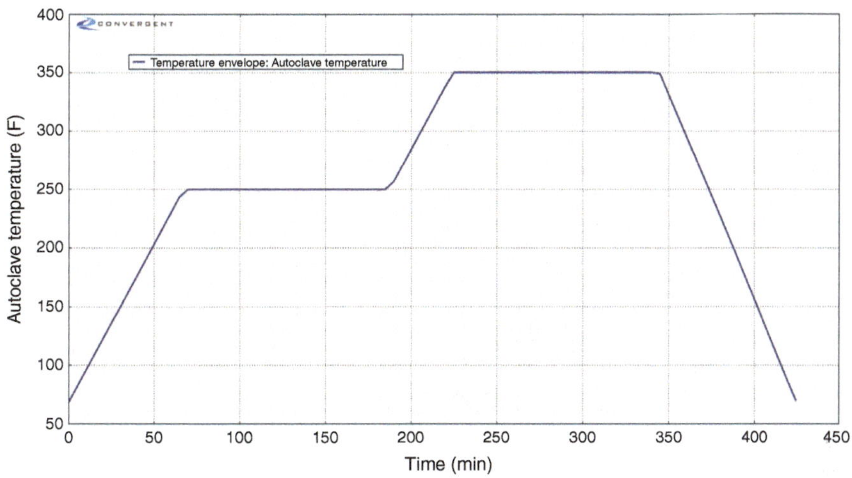

FIGURE 3.22 Cure cycle used for the study.

TABLE 3.5 Temperature distribution at the end of the cure cycle.

Tool	Thermal distribution	Max delta temperature (°F)
Aluminum 606x	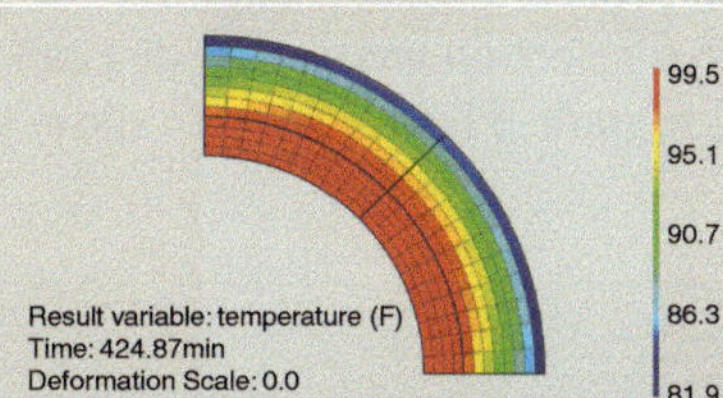	7.9
Copper	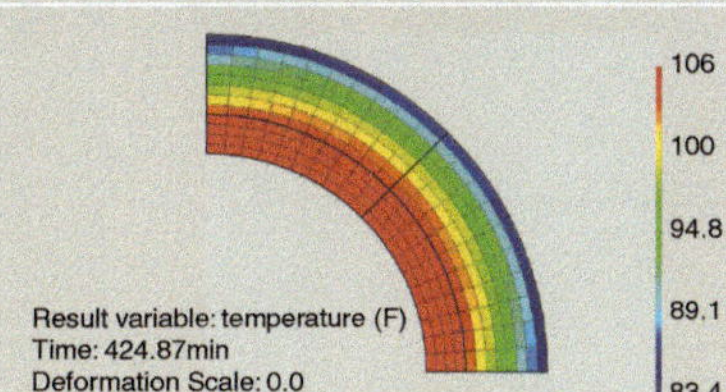	22.6
Invar 32	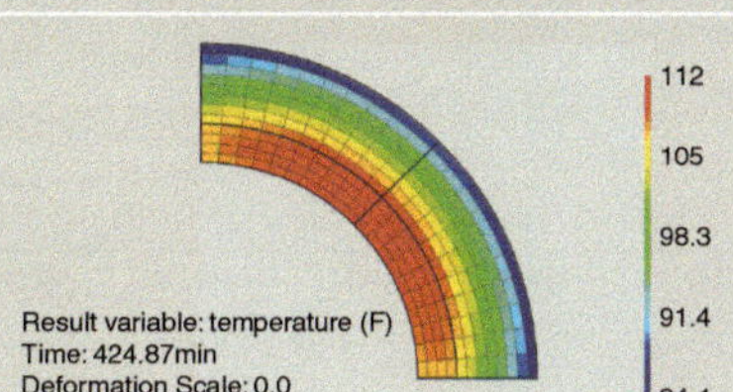	17.6

(Continued)

TABLE 3.5 Temperature distribution at the end of the cure cycle. (*Cont.*)

Tool	Thermal distribution	Max delta temperature (°F)
Steel 1020	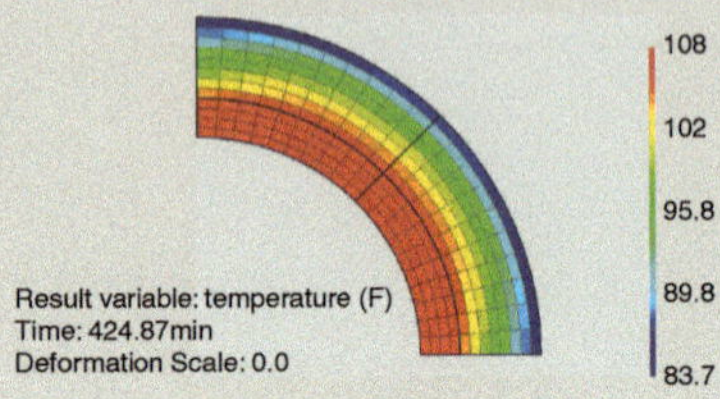	18
Composite	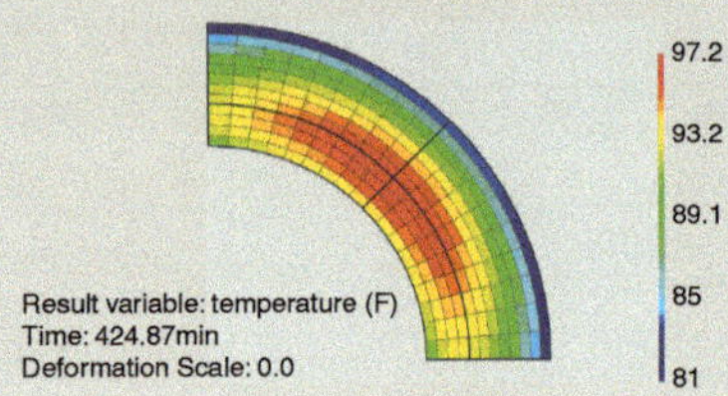	6.2

variation is shown for the copper tool which is 22 °F and that is considered too high if we were to set an acceptable limit of 10 °F max variation. This simple yet effective simulation process provides engineers with a quick way to make decisions on tool material selection and design that can aid during the different design phases of any program. Figs. 3.23–3.27 show the thermal distribution

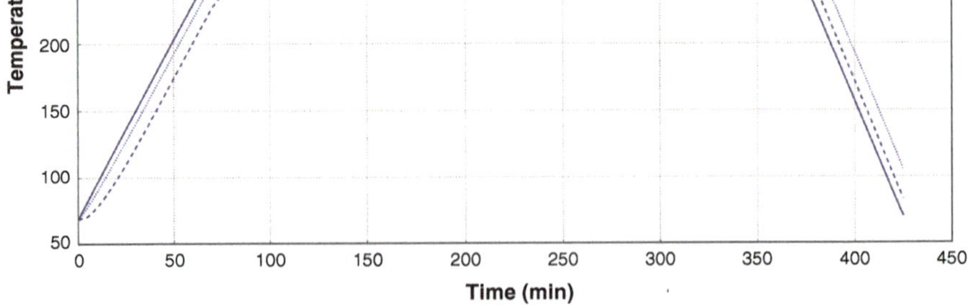

FIGURE 3.23 Part thermal distribution during the cure cycle for steel tool material.

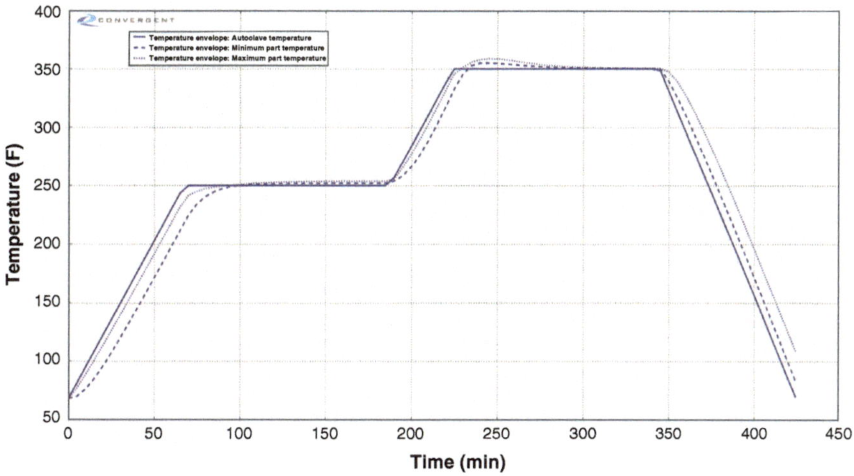

FIGURE 3.24 Part thermal distribution during the cure cycle for invar tool material.

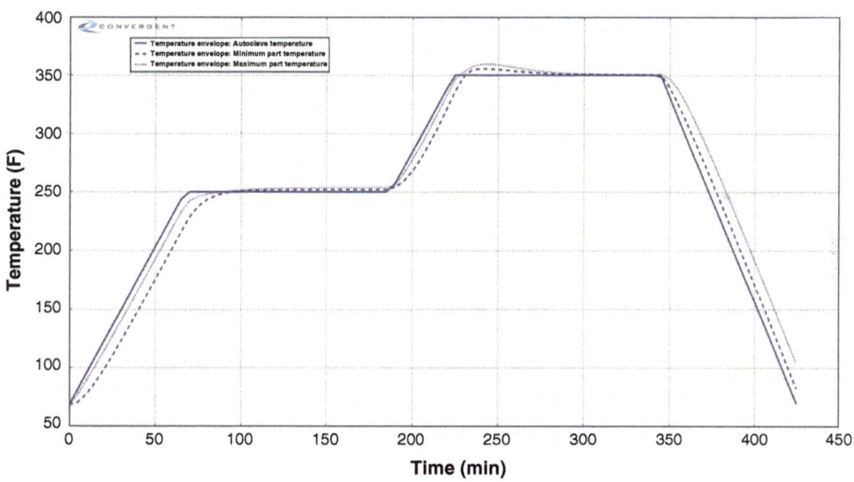

FIGURE 3.25 Part thermal distribution during the cure cycle for copper tool material.

of the part (maximum and minimum temperature) of all tool materials used in this study. These plots offer additional set of data that engineers can use to make modifications to the cure cycle or part design as well to ensure appropriate manufacturing process down the road.

Chapter questions

1 List five common materials used for composite lamination tooling?
2 What are the different bonding joining techniques available?

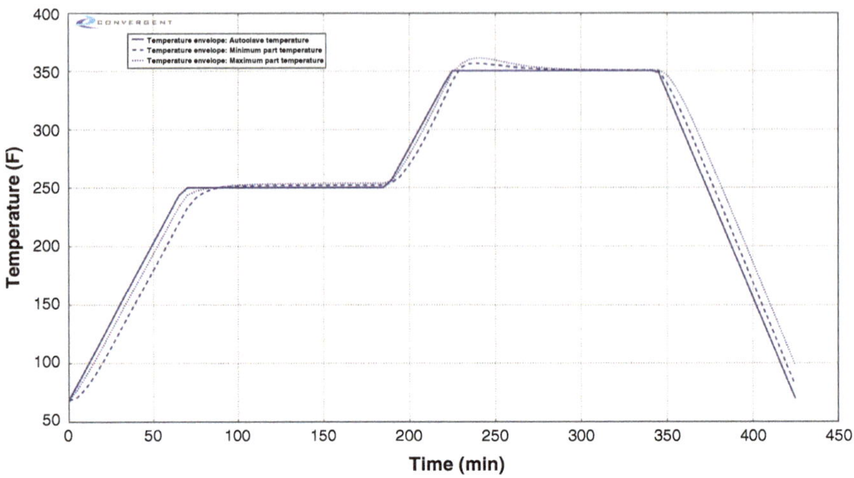

FIGURE 3.26 Part thermal distribution during the cure cycle for aluminum tool material.

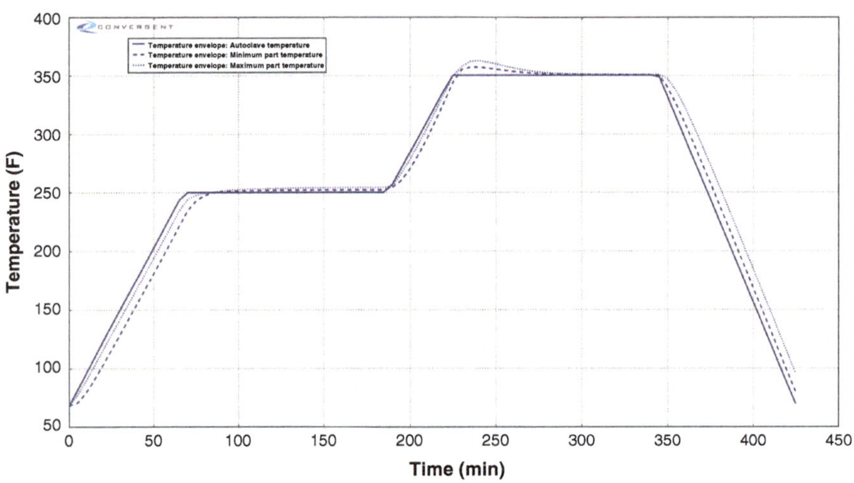

FIGURE 3.27 Part thermal distribution during the cure cycle for composite tool material.

3 Why isn't it recommended to use liquid adhesive form for large bond gaps?
4 What common tooling material has the highest CTE mismatch with composite materials?
5 What steps can a design engineer take to better inform themselves on tool material selection and design?
6 List common resin forms and systems used during composite processing?
7 What are common defects observed during the operation of tooling?

References

[1] http://touchstoneac.com/metal-composite-tooling/.

[2] http://www.ompm.it/composite/.

[3] https://ascentaerospace.com/coastcompositeshome/.

[4] Burden JH. Society of Manufacturing Engineers. Monolithic graphite as a tooling material for carbon fiber composite. Dearborn, Mich: Society of Manufacturing Engineers, 1988.

[5] Burden JH. An Overview of Monolithic Graphite Tooling, In: 34th International SAMPE Symposium and Exhibition, Reno, Nevada, May 8-11. 1989 pp. 1315–1325.

[6] Boatman ES, Covert D, Kalman D, Luchtel D, Omenn G.S. Physical, morphological, and chemical studies of dusts derived from the machining of composite-epoxy materials. Environ Res. 1988; 45 (2):242–255.

[7] Luchtel DL, Martin TR, Boatman ES. Response of the rat lung to respirable fractions of composite fiber-epoxy dusts. Environ Res 1989;48(1.):57–69.

[8] Martin TR, Meyer SW, Luchtel DR. An evaluation of the toxicity of carbon fiber composites for lung cells in vitro and in vivo. Environ Res 1989;49(2):246–61.

[9] https://www.nlr.org/capabilities/composite-repair/composite-repair-3/.

[10] https://www.teslarati.com/spacex-bfr-tent-spy-shot-mars-rocket-tooling-molds/forward-fuselage-of-the-787-on-a-mandrel-credit-boeing/.

[11] Hart-Smith LJ, Redmond G, Davis MJ. The Curse of the Nylon Peel Ply. 41st International SAMPE Symposium, March 24-28, 1996, pp. 303–317.

[12] https://www.eenewseurope.com/design-center/benefits-plasma-cleaning.

[13] https://www.compositesworld.com/articles/plasma-treatment-as-surface-preparation-for-adhesive-bonding.

[14] https://www.antala.uk/surface-preparation-for-adhesive-bonding/.

[15] http://www.reuthermold.com/aerospace.html.

[16] https://www.instructables.com/id/Carbon-Fiber-Aircraft-Flap/.

[17] https://www.convergent.ca/products/raven-simulation-software.

Chapter 4

Design approach and guidelines

Chapter outline

This chapter is devoted to design details for both lamination and assembly tooling. We will provide the reader guidance and recommendations on how to perform the design and requirements that shall be considered during the planning phase. But first, we want to give the reader an overview of how a program that requires some sort of design is organized and where tool design specifically fits in the big picture.

In most aircraft programs the design usually goes through many phases and each phase has its own criteria on what should be evaluated and what should be left to a later date. When considering tool design specifically, it usually cannot start without having part designs finalized or at least drafted in order to understand the overall shape and design of a tool. This is partly why during the early stages of a program no tool design is considered. Fig. 4.1 shows an outline of the typical design phases for any airplane program. The conceptual design phase is the first part of the execution and focuses on general sizing of the aircraft exterior, performing the necessary aerodynamic studies and overall airframe configuration. No drawings or detailed design is created at this stage. Generally speaking, a conceptual design phase can last between 6 months to 2 years in some programs and is a crucial step on setting the stage for all follow-up work.

If the program had sufficient funds and the conceptual phase was deemed successful, that is when the team will endeavor to start the preliminary design phase. In this phase, additional details are added to the design and more work is done on the part level designs and the focus is not only given to the airframe loft anymore. Since additional part details are added at this stage we can also start by working on tooling concepts for the lamination tools and the tools

Tooling for Composite Aerospace Structures. http://dx.doi.org/10.1016/B978-0-12-819957-2.00004-3

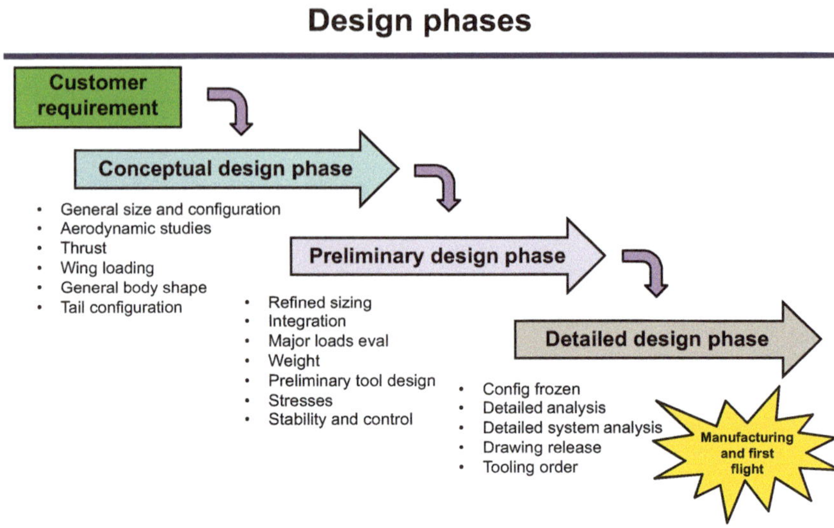

FIGURE 4.1 Design phases for a typical aircraft program.

needed for the assembly of the airframe assuming the assembly sequence has been outlined as part of the study. It is important in this phase to start the work on tooling concepts since the lead time for many of the tools is very long and you want to have some data to use for writing statements of works (SOW) that go out to suppliers to bid on building these tools as needed. The more information you have about the tools and their designs in the SOW, the more accurate the bids you receive back from the suppliers to make sound decisions on how to precede. Even though we expect the part designs to change down the road, we can mitigate some of the risk by asking the tool fabricator to start procuring the material needed and performing any task that does not require fabricating the actual tool until a more firm design is available. This phase of the design can range from 1 to 2 years. A preliminary design review (PDR) is usually the conclusion of this phase and takes the form of a meeting that lasts several days with advocates and subject matter experts (SMEs) to review all the content and provide necessary feedback. A pass-fail criteria is usually set and based on the outcome the program can proceed to the next phase.

During the critical design phase the team will start the work on the detail level and the generation of design drawings for all the parts. The bill of materials (BOMs) will be generated and minimal changes will need to occur post this phase to minimize any risk moving forward. Depending on the project, this is the stage where tool orders are made to the suppliers to start the tool fabrication as the parts are expected to have firm designs and only minimal changes can occur that will not impact the tool. If large changes are made to the part it

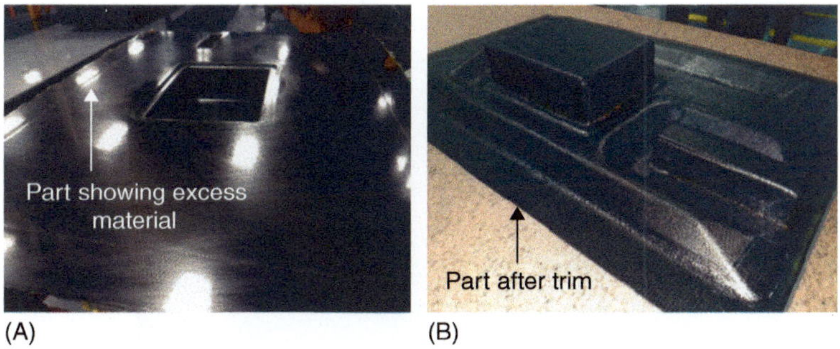

FIGURE 4.2 (A) Part showing the as built laminate post cure and the (B) part after trimming.

can cause the entire tool to be scrapped. This phase lasts the longest from 1 to 3 years. The critical design review (CDR) is where all the details of the design is reviewed that includes the part design, unique features, schedule, risk and mitigation. A pass fail criteria is also set in this case which signals that the program is ready for production or not.

We will mainly focus on composite part designs and associated lamination tools as well as assembly tooling. Prior to getting into the details associated with tool designs we need to have some general definition of tooling concepts that are typically used in this field to help the reader understand the concepts in the remaining chapters.

These definitions can be summarized as follows:

1. End of Laminate (EOL)

Since the manufacturing process of composite materials involves the use of layers (laminas) that are stacked on top of each other to create the part it is inevitable that those layers will extend beyond the point of where the actual part should end per engineering drawing. In order to ease the manufacturing process, the layers are allowed to extend on the tool and are subsequently trimmed post cure. Fig. 4.2 shows a part that has been fully manufactured post cure with the excess material and the part after trim showing the final product that adheres to engineering definition.

2. End of Part (EOP)

This is referring to the location where the actual part ends based on the engineering design definition. As noted above, since laminating a composite part requires the use of layers (laminas) they will extend beyond the boundaries in order to ease the manufacturing process. This excess material will be trimmed after final cure via a CNC machine or manually by hand in order to achieve the final part. Fig. 4.3 shows a lamination tool that has the EOP outline highlighted.

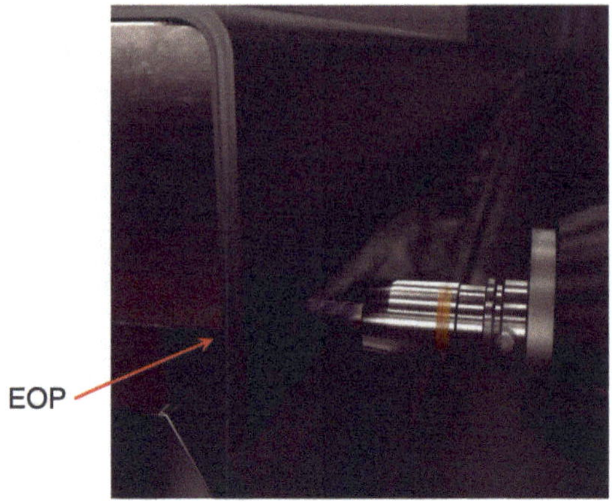

FIGURE 4.3 Lamination tooling showing the end of part highlight in the yellow perimeter [1].

3. Scribe Lines

These are features that are included on the tool in order to aid the manufacturing team during the fabrication processes of a composite part. The lines act as guides to the laminators working on the shop floor on where to terminate the laminate during the fabrication. It shall be noted that these lines need to be minimized on the tool in order to mitigate any confusion that might occur or impact on the structural integrity as they are done by removing material with specific depth from the tool. When including these features the depth of the line is recommended to be between 0.010 and 0.020 inch and include at least two lines around the tool or more as required. Fig. 4.4 shows an example of scribe lines on a tool.

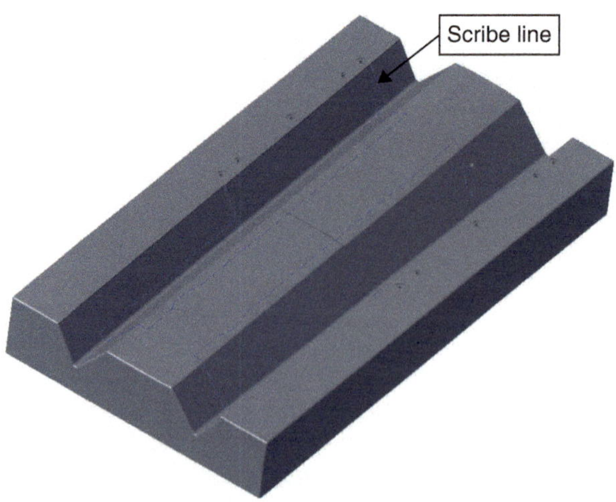

FIGURE 4.4 Lamination tool with scribe lines highlighted.

FIGURE 4.5 Tooling tab used as position feature on the part [2].

4. Positioning features

In order to ensure that all the individual parts fabricated fit the high level assembly without any issues such as excessive gaps or interference, the utilization of coordination holes or tooling tabs can be used which are examples of positioning features that are typically used. These features should be located outside the EOP and can be removed after using them via manual hand trim in the case of tabs as shown in Fig. 4.5.

The tool shall be designed in such a way to allow for these provisions. Hardened drill bushings of specified diameters can also be used to achieve the objective of positioning where they are installed flush to tooled surfaces. The exact location of where they need to be installed shall be based on the engineering requirements and assembly plan.

Another aspect where positioning features are needed is for laser projections (which will be discussed shortly) on the part. Those features are needed in order to help production during the set up of the lasers. They should be placed around the tool periphery located beyond EOP by about 2 inches in order to prevent any interference with vacuum bagging. It is recommended to use quantity of 6 features under a projected area. They are hardened, open-end bushing, installed flush to the surface, designed to accept standard retro-reflective optical targets.

5. Female and male tooling

One of the many different nomenclatures used in tooling design is the term male and female tooling. A schematic diagram of each type is shown in Fig. 4.6. When it comes to designing a tool, selecting the appropriate method is crucial since each approach has its own advantages and disadvantages but it also depends on the application of where it is being used.

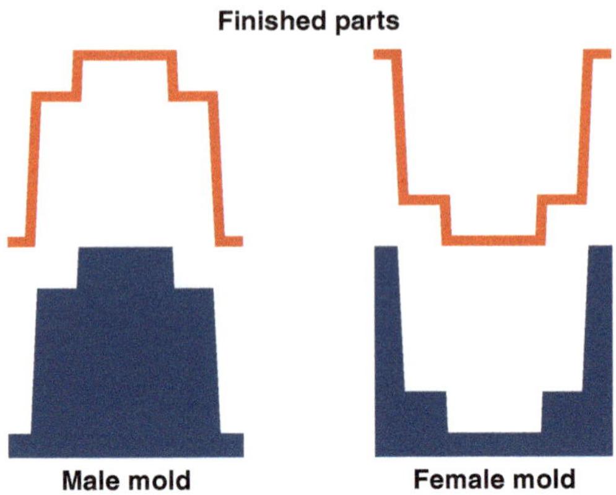

FIGURE 4.6 Male and female tooling.

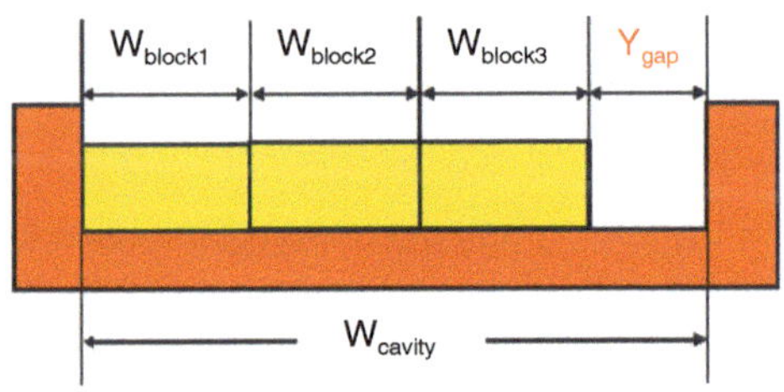

FIGURE 4.7 Schematic diagram showing the concept of tolerance stack-up and how it can add up to create part gaps impacting higher level assembly and installation.

6. Tolerance stack up

It is important for any design to have a tolerance stack-up analysis done and complete prior to moving forward with drawing sign-off and release. Having an incomplete tolerance analysis might cause gaps in higher-level assemblies that might impact fit-up and cause nonconformance's and delays, which impact cost negatively. Fig. 4.7 shows a schematic diagram of how tolerance stack-up can cause the creation of gaps during assembly. If each individual part was manufactured without taking into account the maximum or minimum tolerance allowed per the drawing that can create adverse impact on the assembly. In order to mitigate the impact, the designer needs to ensure that the tolerances used on the drawing is reasonable based on the application and how it will be used during production.

7. Laser projection

The lifecycle of every composite part starts with having a kit of plies that will need to be laminated with a specific stacking sequence as defined by the engineering drawings. As you might imagine, laying each ply on a tool and on top of each other is a fairly hard and time consuming process let alone the need

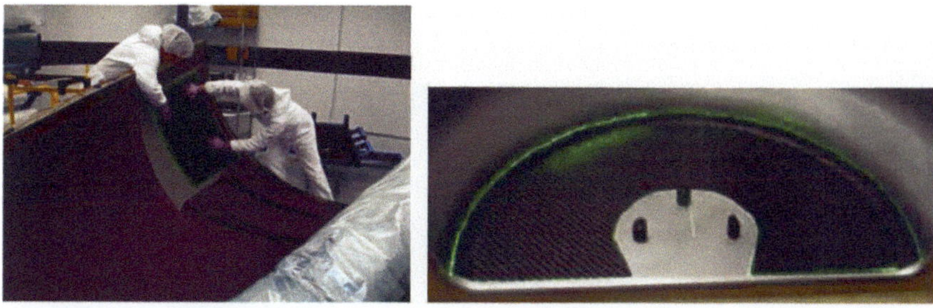

FIGURE 4.8 Laser projection used to simplify the lamination processes during manufacturing [3].

to ensure that every ply adherers to strict engineering requirements set by the process specifications. In order to ease the burden on production, the use of laser projections has been widely adopted over the past two decades. Fig. 4.8 shows the arrangement of how laser is projected onto a tool and how the technicians can then layup each individual ply to match the layout without any need to guess or follow some arbitrary line based on design drawing interpretation.

8. Master molds

The master mold is the support structure used for making a composite tool. The composite tool is then used to build the actual part as shown in Fig. 4.9. The master mold is usually used very few times, often only once which makes the materials considered to fabricate them very different than the final tool. When the master mold is designed, it is important to have the two next steps in mind which are the tool and the final part to ensure the best possible outcome. Since the mold might only be used once, it can be made of a material that is not so

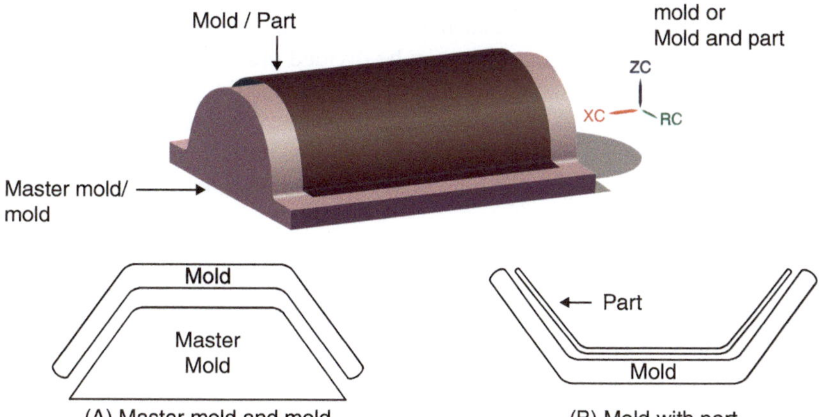

FIGURE 4.9 Master mold example used to build composite tooling.

FIGURE 4.10 Different types of tools used in production (A) assembly tools (B), (C) lamination tools [4–5].

durable and has a lower cost. One of the main differences in choosing material for master molds and tools is the temperature the curing should be carried out in. For typically low temperature components, the curing will be between 200 and 220°F. And since the cure will occur at lower temperatures the material selection is much easier given the wider material options and the prices are typically much lower.

With those definitions now clarified we can focus on aspects related to tool design. Generally speaking, tooling can be divided into two primary types; the first is hot tools that are used to build composite parts and typically see temperatures ranges from 250 to 400°F when cured in an oven or autoclave. The second type is cold tools that are used for either assembly, part trimming, or handling and operate at room temperatures in most instances. Fig. 4.10 shows examples of these types of tools.

When it comes to the design of such tools there are different considerations that need to be addressed in each. This chapter will address those considerations and discuss other aspects of tool design that should be considered by all parties involved from engineering to procurement.

The tool design process in the industry thus far has been mostly based on experience from engineers that have done similar work in the past. There is also very limited documentation or comprehensive writing on how to approach tool design (until now with this book) and its connection with composite manufacturing and assembly. This limits the ability to perform optimization because the tools need to be tested before their performance can be evaluated and optimized [6–8]. Within the past decade advancements in new manufacturing process simulation capabilities allow virtual evaluation within the design phase of the mold, thus enabling more sophisticated optimizations before the mold is actually produced [9–10]. The simulations provide the ability to improve on the thermal response and optimize the design based on the data received which has a direct impact on manufacturing times and quality [11–13]. Chapter 7 of this book will be devoted to discussing this topic and provide an example on how to perform this type of simulation using tools available in most academic institutes and companies.

If the design was not done appropriately, there may be a need for redesigning or reproducing an already existing mold which is expensive and will therefore only be done if the part quality is not within engineering specification. In order to prevent that during the design phase of any tool, there are many considerations that need to be thought-out when designing the right tool for the specific application. Here are a few of the most essential questions that should be asked before starting any tool design:

- How will the tool be used and what performance requirements are required for that use?
- What material best meets those requirements?
- What are the dimensions of the final tool?
- What is your production rate?
- What are the curing conditions?
- What are your tolerance levels?
- How many times does the tool need to perform its operation?
- What contours and integrated functions does the product have?
- What is the required surface finish?
- What is the time frame from prototyping to production?
- What is your budget?

With those question posed, the engineer needs to consult with program management and procurement to coordinate on the best path forward for optimized performance and best success for the program. From the technical and programmatic side you need to consider the requirements that are set by the engineering team. Some of the important requirements that might be set are those mentioned below:

- Coefficient of thermal expansion
- Dimensional accuracy and stability
- Vacuum hold
- Surface finish
- Durability

- Environment, health, and safety
- Weight
- Cost
- Machinability
- Repairability
- Heat and pressure
- Material lifetime
- Maintenance
- Adaptive work on part
- Curing conditions
- Lead time

1 Hot tools

These are the tools that are generally used to laminate composites to take the shape of the as designed part. In order for the composite material to cure it needs to enter either an autoclave or an oven as shown in Fig. 4.11 and that all depends on the material system used and the quality level needed for the part. Autoclaves unlike ovens provide pressure to the part and can aid in removing many of the volatiles that are trapped inside the part creating a part with low porosity content. Some examples of cure cycles used for aerospace material systems are summarized in Fig. 4.12. These cure cycles are largely based on the material manufacturer but should always be studied carefully as modifications might need to occur based on the design. For example, if the part was very thick there is a possibility of having an exothermic reaction in the material and that can cause the part to overshoot the temperature allowed per the process specification as shown in Fig. 4.13. In order to mitigate that from happening one solution could be to add an intermediate dwell to the cure cycle as to prevent that from occurring. The dwell can be at 250°F for 2 hours or depending on the design of the part including material selection and thickness. Fig. 4.14 shows the change in the part temperature when the dwell was introduced and the exothermic reaction was eliminated.

FIGURE 4.11 (A) Autoclave and (B) oven used for curing composite materials [14–15].

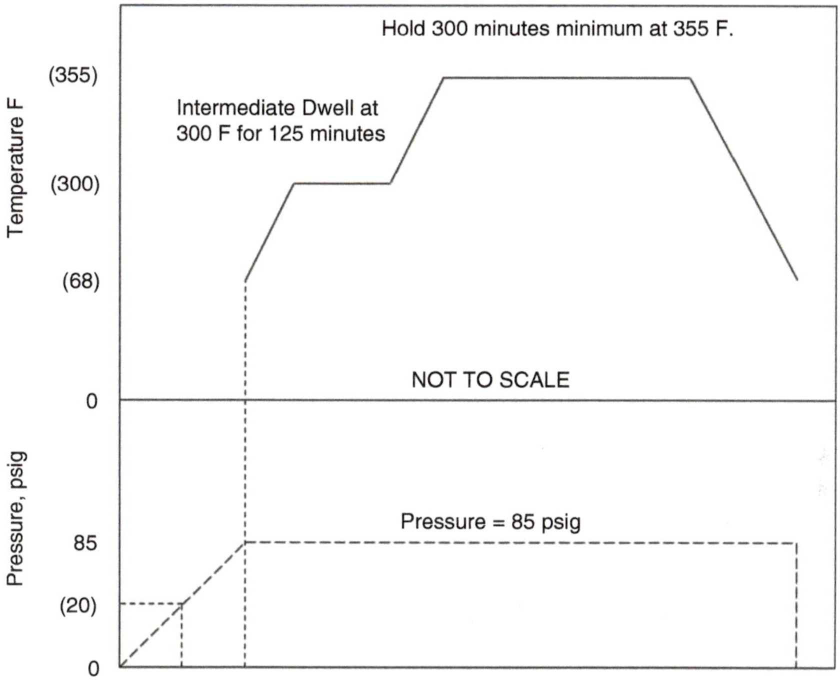

FIGURE 4.12 Cure cycle example used for a typical thermoset composite material.

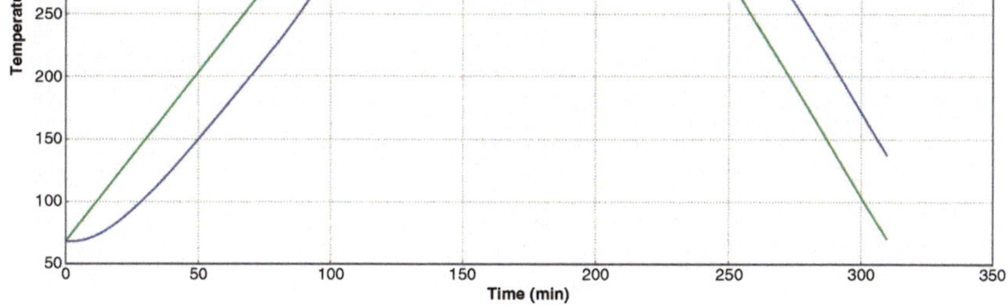

FIGURE 4.13 Material experiencing exothermic reaction due to the cure cycle used for that material and design.

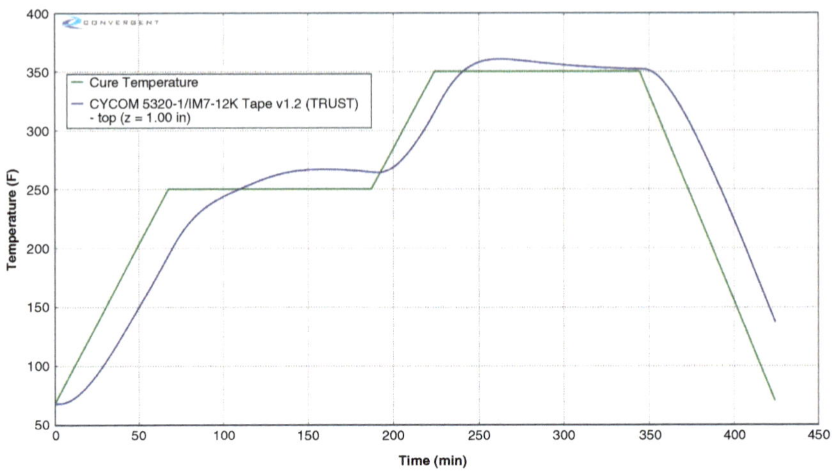

FIGURE 4.14 Cure cycle modification with an intermediate dwell to prevent the exothermic reaction in the part.

1.1 Requirements

When it comes to production parts for larger programs that are not R&D, the lamination molds (LMs) shall be designed and fabricated to withstand a minimum of 50–100 cure cycles without degradation that would cause a detrimental effect on part quality or the dimensional conformity of the end item (e.g., vacuum integrity, warpage, surface defects such as pitting, etc.). This requirement might be difficult to guarantee as it depends on the material you select and the way it is operated. But with time and utilization a data base will be generated in order to validate the integrity of the tool during time.

The vacuum integrity shall be maintained over the life of the tool. Vacuum integrity is defined as less than 1 in-Hg lost per 20 minutes hold for the life of the tool. Final verification shall be proven by a documented leak test performed at the maximum processing temperature of the tool (e.g., 350 °F). LM vacuum-tight performance shall be a minimum of 28 in-Hg (minus 1 in. of Hg for every 1000 feet of elevation above mean sea level (MSL) to a minimum of 25" Hg), using calibrated gauges.

Another important criteria to consider will be the surface finish of the tools. For master molds and lamination mold surfaces, finishes should be 125√RMS or better with profile tolerance not to exceed +/-0.010" in localized areas up to 25 square-feet, nor not to exceed +/-0.020" overall in an unconstrained condition. The reason for requiring such a tight surface roughness is the need to have a smooth surface on the exterior of the part or what is typically known as "tool side" for optimized aerodynamic performance. If the surface was not smooth that will ultimately transfer to the part impacting the airplane performance downstream. Moreover, adjoining segments (multi-piece) tooling shall

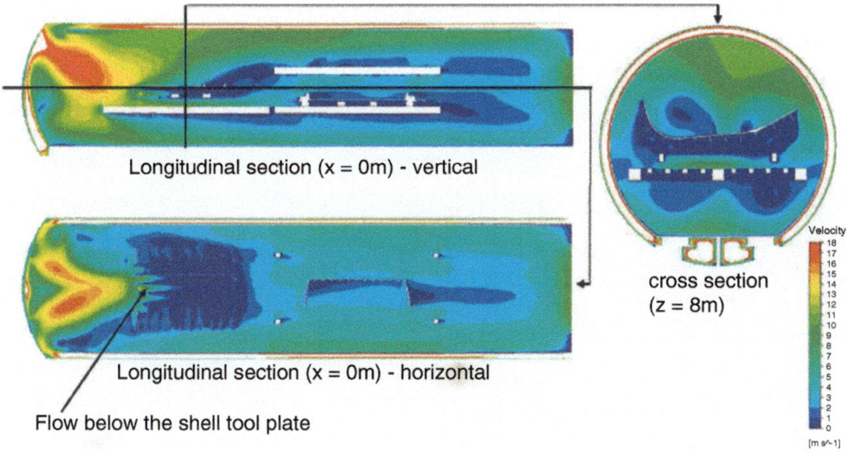

FIGURE 4.15 CDF analysis of autoclave airflow with wing-like structure [16].

have common surfaces aligning with a maximum step of 0.005" in order to minimize areas of mismatch causing anomalies on the part impacting the higher level assemblies.

The coefficient of thermal expansion of the finished tool shall be accounted for during the tool design such that the cured part conforms to the design intent as specified on the face of the drawing per engineering. The process of compensating for the CTE can be done via analysis or trial and error testing. Later in chapter 7 we will be discussing this in more detail, and it is recommended that all designers take this concept into consideration during the design phase and involve the manufacturing and materials teams for input. If this issue was overlooked it can cause the part to fail the dimensional inspections and cause the tool to be re-worked or possibly scrapped.

The tools shall also be designed to promote airflow through the backing structure in order to allow temperature uniformity during cure as much as possible. Some analysis is recommended to be done via computational fluid dynamics (CFD) in order to determine the airflow inside the autoclave and base the design on the results obtained. This analysis depends on the type of autoclave being used as well as material type, size, and number of parts that will fill the autoclave at each load. Fig. 4.15 shows an example of a CFD analysis for an autoclave with a wing skin tool. This will support our understanding of how the tool/autoclave shall be designed to ensure uniformity.

The physical dimensions and weight of the tool is another very large concern when it comes to manufacturers due to the need to move these tools around the shop on a regular basis and considering any special provisions that might cause an issue for the facility. As a general rule of thumb, tools shall not exceed 90 inches in width, 60 inches in height, and 450 inches in length. Weight shall not exceed 6000 pounds. Of course this will all depend on the need and the part

FIGURE 4.16 Large invar tool for wing skin [17].

that is being manufactured. An example of a very large invar tool is shown in Fig. 4.16, in the figure you can see a crane that is used to aid in the movement of that tool around the shop. This specific size of tooling can be handled by a finite set of manufacturers worldwide and requires a very stable infrastructure to support the weight and size.

The backing structure is typically used to increase the strength of the composite mold/tool during transportation, layup, debulking, and thermal cycling. Backing structures can be a significant contribution to the investment in a composite mold/tool. In some cases, customers have quoted up to 50% of the total cost for the backing structure. An example of a backing structure is shown in Fig. 4.17. Key considerations when designing tooling backing structures include:

1. *Tolerances*: determine the demand based on the final part design and what contribution can be expected from the mold/tool.
2. *Environment*: identify what processing conditions the mold/ tool will be subject to (e.g., ambient, oven, or autoclave).
3. *Matching CTE*: the backing structure should match as closely as possible to the tool if there is a mechanical connection between the two. A mismatch in CTE will produce thermal distortion as the mold tool is subject to changes in temperature and could potentially cause inaccuracies in the final cured part.
4. *Attitude*: determine how the mold tool is held during operation of its primary function. For example, a simple shop floor tool fixed to a mobile cart would remain on the ground and horizontal. On the other hand a large automatic

FIGURE 4.17 Tool backing structure example made from composite material with aluminum core [18].

tape layup (ATL) tool requires rotation through a number of axes in conjunction with the robotic deposition head.

5. *Airflow*: evaluate the autoclave and/or oven thermal flow and design the backing structure to aid in creating an even heat distribution.
6. *Connection*: consider the interface between the mold/tool shell and its backing structure. If mismatched materials are to be used, then movement between the two structures must be allowed using flexible adhesives or slotted mechanical fixings (typically on metallic backing structures).

These considerations offer the best opportunity to match the physical properties of the mold and utilize standard products readily available in the market.

Design engineers should minimize complicating tool designs especially in creating loose details. In cases where the tool does include loose or removable details, those must be tethered by cables except when the details are too large, heavy or interferes with the production use of the tool. Loose details not attached to the tool must be placed/located in a "shadow box" that also serves as a storage provision for future "on-demand" use.

Lamination mold tooling solutions may include provisions for integrated vacuum line sources spaced periodically along the periphery outside of EOP, at least every 3–5 feet. Integrated vacuum ports shall be fitted in a manner that prevents damage during tool movement and placement into curing equipment. Each port shall be marked with clear identification for which vacuum source site around the tool perimeter that it corresponds to.

And most importantly all tools shall address all ergonomic and safety concerns. Construction shall avoid unnecessary sharp corners or edges, protrusions that create tripping hazards or head impacts. Platforms or integrated cart surfaces intended for standing or walking shall have anti-skid construction or hi-temp capable treatment. Hand rails, hand guards, and handles shall be appropriately included to facilitate manual movements. Pinch-points or any moveable elements shall have safety features designed to avoid possible injuries.

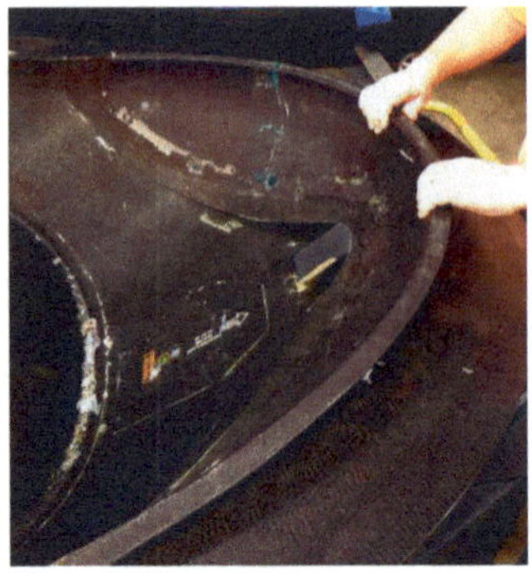

FIGURE 4.18 Example of a removable draft lock feature on a composite lamination tool.

One unique aspect in some tool designs is having removable (draft locked) tool pieces as shown in Fig. 4.18. The purpose of these removable parts is to aid in demolding the part post cure and prevents the part from getting trapped. If such a feature is used in the design, it is recommended to follow some of these suggestions:

1. Include scribe line at the edge of the feature as defined by the native CAD model files +/- 0.030".
2. Add a minimum of two hardened bushings per segment, bonded flush with the LM surface, designed to accept standard 0.250 inch tool pins at mating location defined by the native CAD model files +/- 0.002 inch. Patterns shall be in a manner that prevents misorienting the removable pieces when installed. Bushings shall be blind, and installed so as to maintain vacuum performance for the life of the tool.
3. All removable items shall be identified with item numbers (sequential), tool number, revision and date of manufacture in a location defined in the tool drawing.

As far as other handling features that need to be considered during the design of these tools they should include:

1. *Off-loading*: for tools that have a 500 lb weight. It is recommended to have integral lifting provisions for straps or eye-hooks. Tools of greater than 500 lbs shall be equipped with forklift tubes welded or fastened in place, centered about the finished tool center of gravity. Minimum interior tube size should be based on stress analysis. Fork tube spacing shall be 32–48 inches on center. Tools shall be equipped with marked tie-down points or tie-down rings for use during transportation.
2. *Casters*: Tools of greater than 100 lbs shall be equipped with removable or replaceable casters to facilitate tool movement. Casters shall be of the locking type. Casters shall be compatible with maximum cure temperature (e.g. 400 °F) and of an applicable material.

3. *Lift rings*: Tools shall be equipped with 3 (under 500 lbs) or 4 (500 lbs and over) removable lift rings to facilitate tool movement. Lift rings attachment points shall be equally spaced from tool center of gravity. Lift rings and attachment points shall be designed for expected maximum load with factor of safety of 3.0 or greater. Expected maximum weight should include total tool weight plus part weight plus 10% as a safety precaution.
4. *Leveling feet*: Tools shall be equipped with manually adjustable leveling feet with a minimum of 2 inches of useful travel.

1.2 Quality control

The design should have provisions on the drawing that allow the inspectors on the shop floor or a supplier be able to perform inspections and audits in an easy and straight forward manner. A quality plan needs to always be in place with the support of several disciplines (design, manufacturing and quality) in order to ensure a successful buy off of the tool first time around minimizing the errors with first pass quality. An example of such tool drawing is shown in Fig. 4.19. The drawing highlights several critical design details that can be used by inspectors for checking the tool. It is recommended for every tool to create a tool proof inspection report. The objective of this report is to verify that all dimensions and measurable requirements, drawing notes, and critical features meet requirements. Information pertaining to the tool-use shall also be included as evidence of conformity to the pertinent tool specification.

Another aspect related to quality control is the need for a tool thermal profile. The thermal profile is done to ensure that the tool can reach the temperature required without significant temperature delays due to its thermal mass or design. It also ensures thermal uniformity across the tool prior to it entering service. Having non-uniform heating can cause the composite part from lagging behind requiring the need to slowly heat the autoclave or oven in order to allow the lead and lag thermocouples (TCs) from adhering the specification requirements. If the lagging TC was significantly lagging behind the leading TC that will force the operator or program to reduce the heat up rate to allow sufficient time for it to get up to temperature which can have a big impact on run time that will also impact production rate in the long run.

Lead and lag thermocouples are those identified during a part thermal profile and are the primary TCs that are monitored during a cure and drive the autoclave program in terms of heat up rate and cool down.

A tool thermal profile is done by including a large number of TCs all across the tool as shown in Fig. 4.20 and entering the part into a simulated cure cycle that represents the actual cure it will be operating in. By monitoring all the TCs on the tool we will be able to identify the heat variation in the tool and the need to perform any modification. Depending on the process specification all temperature variations need to adhere to specific requirements. The industry-accepted standard allows for a max temperature variation between 15 and 25 °F

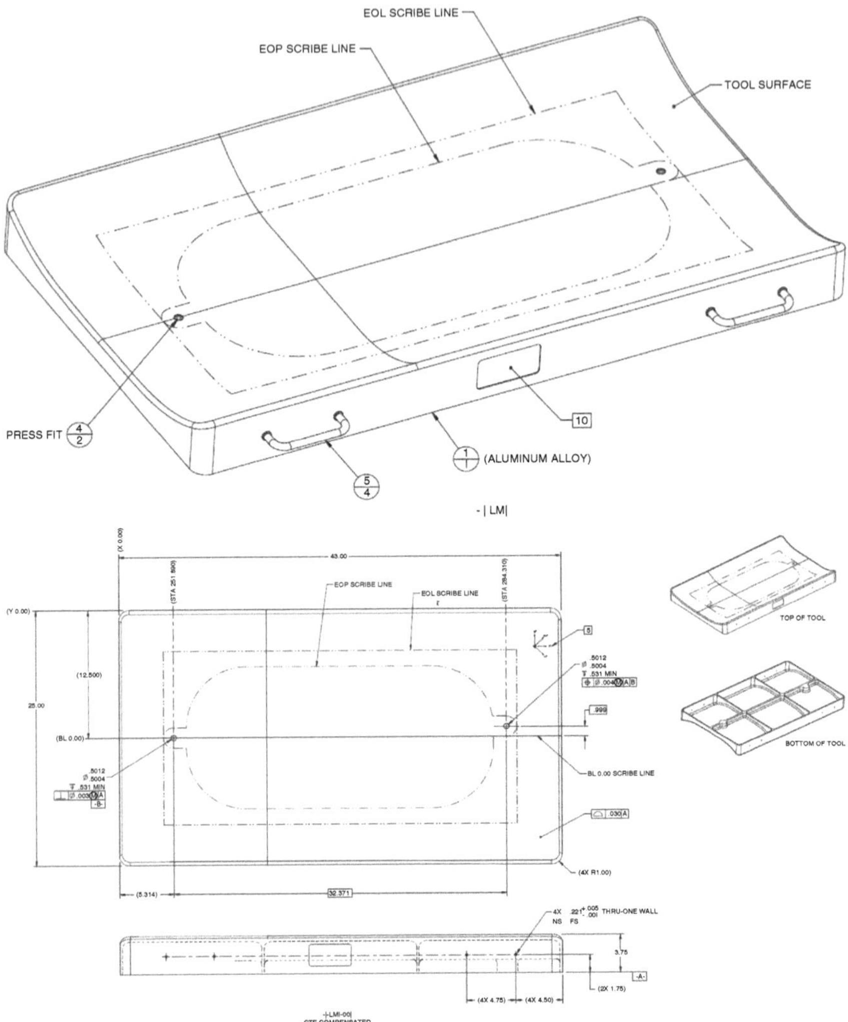

FIGURE 4.19 Example of a 2D design drawing for a lamination tool.

between the lead and lag TC. If the temperature exceeds that it is considered a nonconformance and engineering needs to evaluate and provide a disposition with substantiation on a path forward. Substantiation can be in the form of mechanical testing to ensure no impact on the material allowables and/or physical testing to make sure the part reached the appropriate degree of cure.

Example results from the thermal profile are shown in Fig. 4.21. As observed, at certain locations along the cure the delta temperature between the lead and lag was around 57 degrees, which is a reason to reject the tool until a solution is done whether it be a design change or accepting the tool at risk.

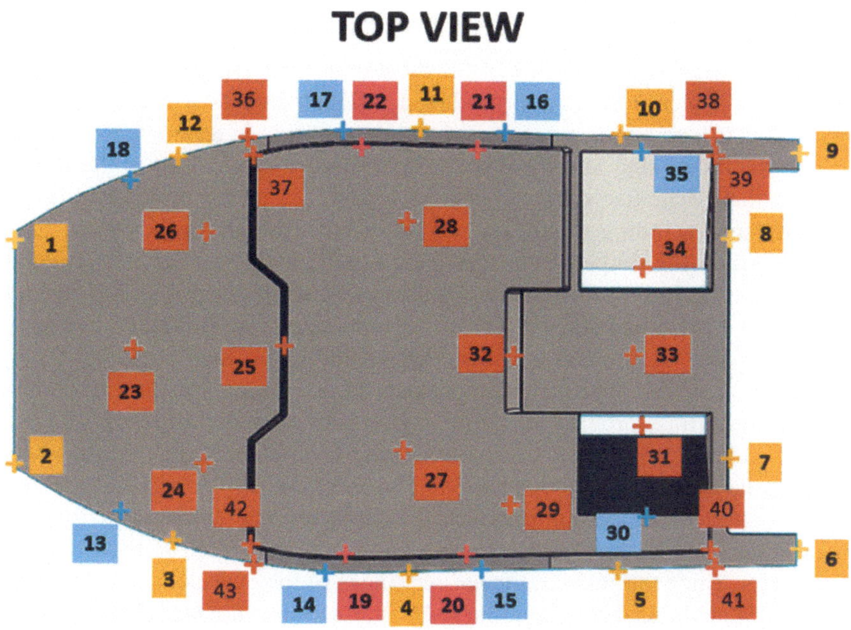

FIGURE 4.20 Thermocouples located across the tool during a thermal profile study.

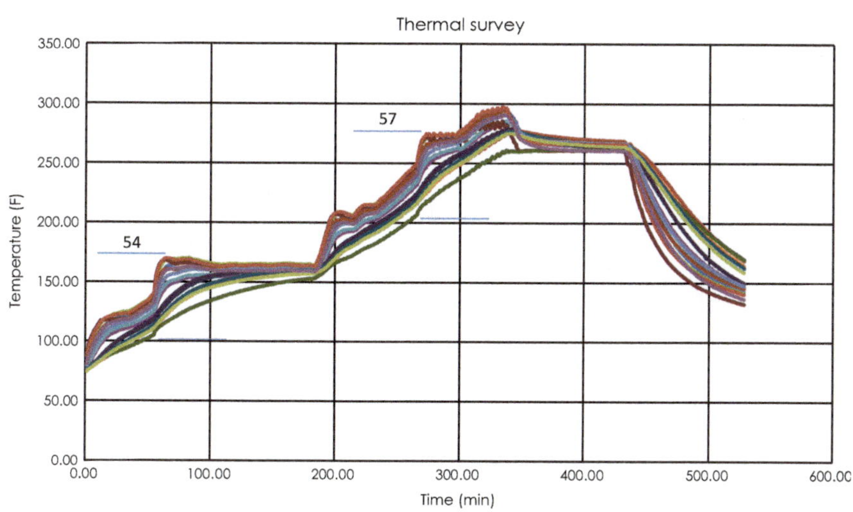

FIGURE 4.21 Results of the thermal profile showing temperature variation based on TC location on the tool.

1.3 Residual stress

When it comes to hot tools one of the most important aspects that a designer needs to consider is the process induced residual stresses and deformations that occur in composite structures during cure. This is an inevitable effect during the processing and there are several studies carried out in the literature on this particular subject [19]. As a designer, it is important to have a better understanding of the process induced residual stresses since they directly affect the shape of the final part which are critical for dimensional tolerances. There are various mechanisms that are responsible for the development of residual stresses and distortions including thermal anisotropy in the part and tool, chemical shrinkage of the resin, tool/part interaction, resin flow, consolidation and compaction, fiber volume fraction gradients, moisture swelling, prepreg variability, gradients in temperature and the degree of cure or crystallization. These have all been identified as mechanisms responsible for process induced residual stresses. When the part and tool are forced together by a certain pressure and subjected to a temperature ramp, a shear interaction occurs between them due to the mismatch in their respective CTEs. Since the resin at the earlier stages of the cure does not have any significant modulus developed, the shear interaction is not significant. As the cure progresses the shear stresses between the part and tool at the interface increases and non-uniform stress distribution takes place causing bending moments upon removal of the composite part from the tooling leading to shape distortions such as warpage in the composite part. Fig. 4.22 illustrates this type of distortion. More detail on his phenomena will be discussed in detail in Chapter 5 and 7.

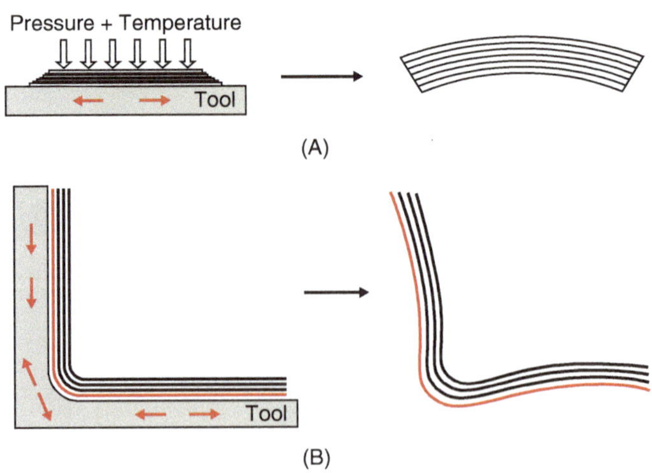

FIGURE 4.22 Composite distortion due to cure of a flat plate and L-shape composite.

1.4 Tool design best practices

Any design will ultimately depend on many variables and some can be out of the hand of the designer due to concerns such as cost, complexity, manufacturing and many other factors. With that said, there is always some tips and best practices that can be followed in order to be on the right path to achieve best design possible. Some of these best practices are noted below in no particular order:

- For composite tooling, ensure that the laminate used is symmetric and balanced in order to minimize any distortion caused by the stacking sequence.
- When possible, the design shall be done in a way that can accommodate design changes associated with inevitable modifications to part designs.
- Evaluate the potential residual stresses that are caused by the composite part in order to compensate for any spring-back or warpage that might be generated.
- Discuss the manufacturing capability of the supplier that will be fabricating the tool in order to avoid adding complexities to the design that are not manufacturable.
- Avoid including tolerance on the face of the drawings that cannot be realistically achieved. This will depend on the tool type and material used. Consult with the broader engineering community on the appropriate values to use.
- Utilize the correct material for the correct application. This decision shall be in coordination with the manufacturing and materials team at the company. For example, avoid using Invar as a material if the shop floor cannot handle the transportation due to the weight.
- Consider the number of cure cycles the tool will experience prior to selecting the material and the design. For small cycles (less than 5), the tools are considered prototype and non-standard or development materials (e.g., tooling board) should be considered which will allow for low-volume manufacturing and use of low-temperature cure material. For intermediate cycles (5 – 20) these are not necessarily considered full production and can use non-standard materials based on the application. For high cycle (50 – 500) these are classified as production tooling and will generally require tools to be autoclave cured and the use of established materials.
- Critical details of the tool shall be clearly identified on the drawing to aid in the inspection and quality control. These critical details should focus on aspects of the process and tool that have a direct bearing on part performance.
- Determine and understand the effects of damage on tool performance and repair process. Depending on the design and material, the repair might require extensive delays and you are better off making certain changes that support ease of repair (design for repair).
- Ensure that the appropriate surface finish is identified on the drawing especially for metallic tooling as any surface cracks might cause fatigue issues in the long term impacting the tool performance.

- Large thermal mass to heat up and cool down can be problematic with larger or thicker parts. Depending on the part and material, appropriate thickness shall be used.
- The support structure or supplied cart for any tool shall be designed in accordance to the needs of the manufacturing team and handling equipment. Stress analysis on the support structure is also necessary.
- For tools made from tooling board, it is recommended it be designed using the standard board sizes available to prevent unnecessary bondlines or bondlines that are close to the final surface.
- All joints between adjacent blocks should be staggered by a minimum of 2 inches wherever possible.

2 Cold tools

When it comes to tools that are used for assembly fixtures, different considerations need to be addressed compared to lamination tools given the different environments that each product operates in. Most of the tools are typically made from aluminum but steel can be another option when aluminum is not a viable option especially in instances when the positional tolerances needed over greater distances is tight. These types of tools need to be designed to maximize rigidity and to prevent distortion in the parts from forces during locating, clamping, drilling, riveting, and fastener installation activities. Dimensional stability is an important criteria when it comes to this type of tooling as any movement during installation or fixturing can lead to inadequate tolerances impacting the assembly process. The use of force to mate the parts is not recommended given the amount of structural damage it might cause reducing the structural integrity of the part.

An important aspect to focus on with these tools when it comes to the design is minimizing the number of pieces for each assembly tool in order to minimize the need for welding at joints. And if welds were required they need to be made per applicable American Welding Society (AWS) specification. Stress relief should be done per SAE-AMS-H-6875 as required during the fabrication process and before machining final features. Provisions should be made for safe placement of larger removable details, and lanyards for smaller items shall be provided. Fastened or permanently bolted joints shall be sub-flush on datum features or part contacting surfaces to prevent mark-offs.

Many of these types of tools include vacuum suction features for gripping, securing and contour control as shown in Fig. 4.23 and they need to be designed in such a way to ensure that they do not apply too much pressure in certain areas compared to others and impact the structural integrity of the part. The tools should have provisions for a permanent mounted vacuum gauge reading in Inches of Mercury and the integrity should be calibrated to not drop more than 1 in-Hg in 5 minutes performed at ambient conditions.

The surface finish for these tools on all surfaces interfacing with the detailed part does not need to be similar to that on the lamination molds and a value of

FIGURE 4.23 Assembly fixture tool with integrated vacuum suction cups.

63√ RMS is acceptable. The removable elements in these tools should have a minimum of two locating features consisting of hardened bushing holes and precision alignment pins. All locating features enable accurate and repeatable location of the tool element and part into the tool in such a manner that cannot be inadvertently reversed, inverted or mislocated. The tolerance of how accurate these holes need to be can be debated but for aerospace structures it is recommended to be within 0.014 inch or better.

As far as the size of the tool and depending on the utilization, it is recommended that the standard working height for all manually performed operations and tool use be limited to 50 inch above the floor or platform where practical with a minimum height of 30 inches. Each tool assembly can be mounted to the floor or engage with floor-mounted pedestals. Each tool should be designed for maximum accessibility to position, locate, and to clamp details, sub-assemblies, master gauges, etc. It is necessary to provide clearance for handling equipment, equipment needed to perform the necessary functions of drilling, trimming, riveting, installing fasteners, and any other processing steps. Adequate clearances are required to easily unload the completed assembly from the tooling. Weight should not exceed 7.5 tons for solo-sling pickup, nor shall it exceed 15 tons for dual-sling pickup that has a minimum 20 ft spacing between lift lines.

Many of these tools also include clamping provisions; when possible, position the clamps to be able to be actuated from the operator's side of the tooling and ensure that the clamp is capable of applying a clamping force that is adjustable. Position the clamps to swing or slide away and not interfere with the loading or unloading of the production part. "Cleco" clamps as shown in Fig. 4.24

FIGURE 4.24 Cleco's used to hold different parts together prior to final installation [20].

should also be considered as an option when designing parts as they are capable of holding a part in place with specific amount of pre-defined force applied that can be controlled and computed.

It is imperative to have sufficient clearance in all areas of the tool to accommodate access required for locating/spotting, drilling, and installation of mechanical fasteners as prescribed by the part assembly process plan. Whenever practical, provide sufficient clearance for the movement and positioning of a robotic end-effector head/spindle during milling, drilling, trimming and probing operations. This provides provisions to expand to use automation for the assembly later on. All assembly joining areas shall at a minimum provide clearance for positioning of manually-operated pneumatic devices for machining, drilling and fastening.

Assembly fixtures should be designed to require the minimum number of removable details to load, locate, position, clamp and hold detail parts into the fixture to reduce complexity and potential errors. The preferred method where required is a pinned feature where the tool detail can be swung out of position and re-pinned, clamped or otherwise precisely fixed into position to facilitate part location. Hardened, abrasion-resistant mounting surfaces shall be provided for support and alignment of tool details as well as the establishment of tooling reference planes. Use of T-bolts, bullet nose dowels, diamond pins, and standard commercial-off-the-shelf (COTS) tooling components is desired. For large and bulky removable tool details, an expanded metal shelf or suitable rack shall be positioned on the tool or on a compact handling cart. Locators are not to be removable from the tooling unless it is the only practical solution.

The use of shims to fill in gaps is something that has been done throughout history but given the advances and complexity in designs we try and avoid them at all cost given the complexity they create in the production assembly phase. Fig. 4.25 shows an example of 2 parts and a tool used to measure the

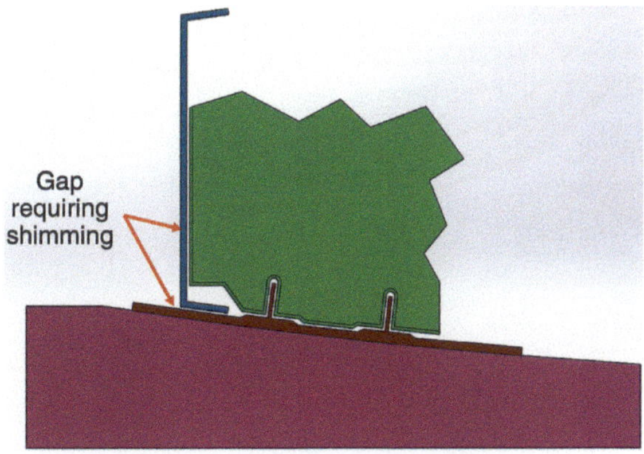

FIGURE 4.25 Gap between 2 parts that will need to have a shim [21].

gap in-between that will ultimately need to be filled weather by a liquid or hard shim. The use of shims to achieve design location and positional accuracy should be minimized and shall not exceed .050" unless approved by engineering. Provisions or allowances for "liquid shimming" to fill gaps requires engineering approval and shall be included in the tool design and fabrication record. All hard shimmed interfaces shall be affixed by pinning or use of torque-stripped fasteners.

2.1 Assembly

Traditional aerospace assembly solutions rely on getting individual aircraft components and manually locating and constraining them using large monolithic structures called assembly fixtures or jigs. Some design aspects for how to build these tools have been discussed in the previous section but what about the assembly process itself. Does that have any impact on how we approach the design?

Consider a wing for a moment such as that shown in Fig. 4.26, the assembly procedure of that structure relies highly on the construction of the build-philosophy, but a common approach for the process is described by [22] as four levels of assembly. First, individual parts of the wing box such as skins, stingers, fasteners are premanufactured. This is followed by the panels and ribs which are assembled by creating a pattern required to provide integrity of the wing box. The panels are then installed on the patterns to create a wing box. Finally, multiple wing boxes are assembled to each other to form the complete wing structure. A flow chart of this process is shown in Fig. 4.27.

As you can tell this is a lengthy process and requires tools that are accurate in order to achieve a successful assembly process and minimize errors and reworks which should be avoided. This all starts with an appropriate design of the tool with a clear plan for assembly. A different philosophy for the assembly

FIGURE 4.26 Example of wing structure [23].

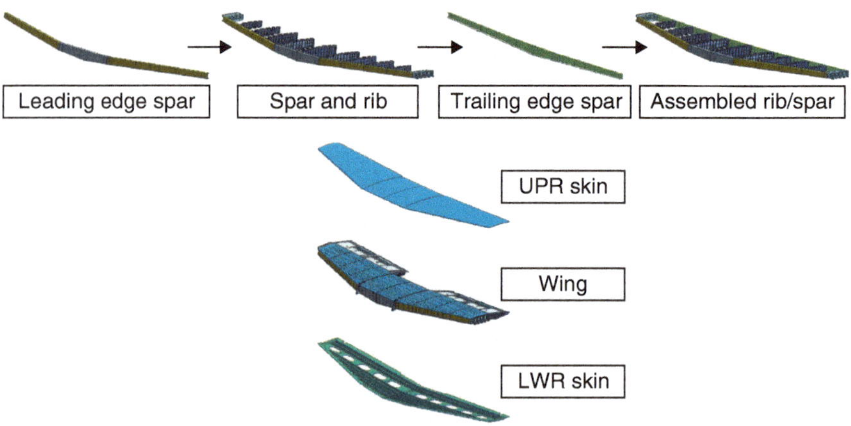

FIGURE 4.27 Typical wing assembly sequence.

can come in the form of utilizing sub-assemblies on the smaller scale and then joining those different sub-assemblies into the final product. This allows for discrete assembly stations and simultaneous ongoing work rather than the need to wait for a specific operation to complete before starting the next. One disadvantage is the need for additional tooling for each station. This should be evaluated and considered during the planning phase for the assembly approach by the engineering team. An example showing the difference in this approaches is shown in Fig. 4.28.

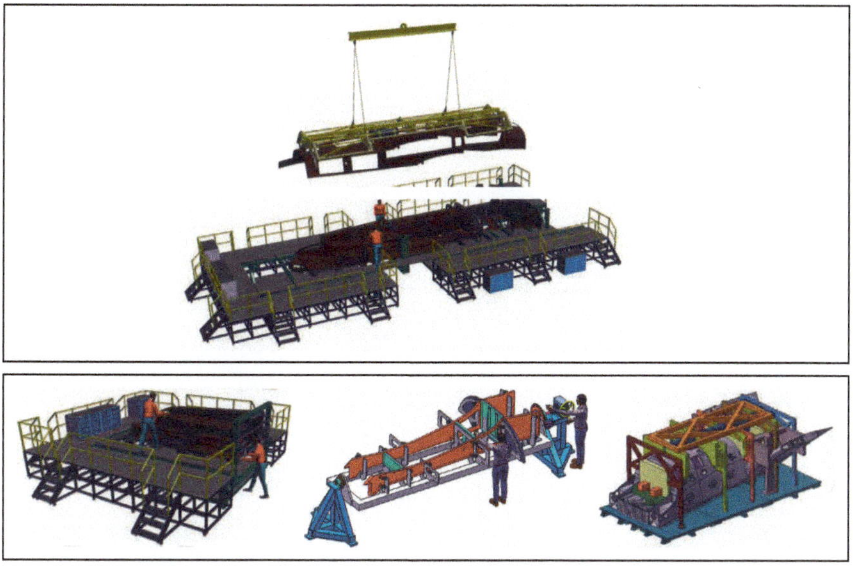

FIGURE 4.28 Full assembly process versus sub-assembly.

These assembly fixtures and jigs are expensive to manufacture and generally offer little or no adjustment at all to accommodate design changes or product variants later on which can have a very large impact on program schedule and cost. Additionally, there is no real time indication of the structure condition and it is not uncommon for an aerospace assembly fixture to fall out of tolerance, causing assembly errors which are passed downstream. Unfortunately this type of error cannot be captured easily and not until the product inspection that these issues are detected and identified, causing product and assembly post-processing and increasing both the cost and lead-time of the product. A relatively new concept called Measurement Assisted Assembly [24] has been proposed for the modernization of aerospace assembly processes, that is, improving their efficiency while reducing the manufacturing costs. Some of the advantages it offers is better positioning accuracy of the components and a significant reduction of the rectification and rework requirements that are usually common with traditional assembly processes. It was shown that a positioning accuracy of 0.004 inches can be achieved for large airframe components using this technique. A concept that tool designers might need to study in order to determine the best path forward for the appropriate approach.

The design of these tools does not require many considerations for temperature expect for those used as bonding jigs that enter an oven needed to cure an adhesive bond for example. These tools do not have any impact on the part manufacturing and hence, the drawings associated with these tools should be simpler than those used for lamination tools. An example 2D drawing is shown in Fig. 4.29.

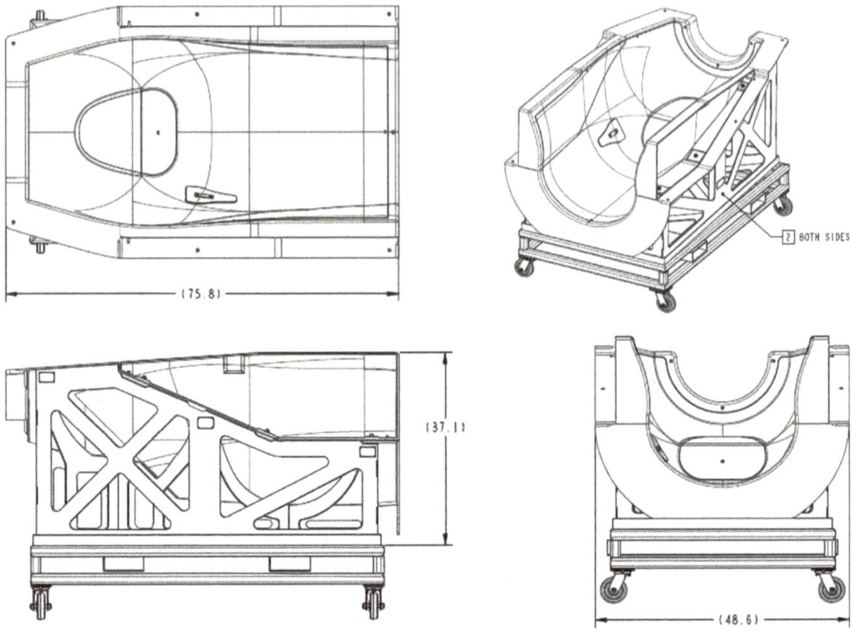

FIGURE 4.29 Example of an assembly fixture tool design.

Another concept for the tool design is the use of flexible tooling which has been a focal point of investigation for various industries with an emphasis on cost and time effectiveness. The flexibility requirement in combination with general requirements of tooling such as rigidness and repeatability can be a very overwhelming task. A LOCOMACHS (Low Cost Manufacturing and Assembly of Composite and Hybrid Structures) project created an automated flexible tooling to meet the demands of future aerospace production to overcome some of the difficulties associated with flexible tooling [25]. The work focused on a creation of a tooling technology that can facilitate the process requirements of an automated wing-assembly by using flexible tooling and intelligence support from a force sensor. These types of assembly tooling can be evaluated by the design team to select the most effective path forward to achieve the ultimate goal of assembling the airframe with minimal setback within the set budget and schedule.

Chapter questions

1. Name five considerations that engineers should account for during the design phase of lamination molds?
2. Name three differences between hot and cold tools?
3. Define EOP?
4. What are the different phases that an airplane design goes through and define details of each step?

5. What is the purpose of a thermal survey for lamination tools?
6. Define the importance of a lead and lag thermocouple?
7. What are the two types of equipment used to cure a composite part and what is the difference between both?

References

[1] https://www.pcminnovation.com/en/molds-for-composites/.
[2] https://www.fictiv.com/hwg/assemble/how-to-design-assemblies-to-be-self-locating-and-self-fixturing.
[3] http://www.3dgage.com/laserprojection.html.
[4] http://www.aint.com/tooling/capabilities.
[5] https://www.sciencedirect.com/science/article/pii/S0034361711700753.
[6] Weber T, Balver, JM. Manufacturing process simulation for tooling. optimization: Reduction of quality issues during autoclave manufacturing of composite parts. Paper presented at the SAMPE Europe Conference, Amiens, France, 2015.
[7] Weber TA, Arent JC, Steffens L, Balvers JM, Duhovic M. Thermal optimization of composite autoclave molds using the shift factor approach for boundary condition estimation. Manuscript submitted for publication, 2016.
[8] Weber TA, Arent JC, Münch L, Duhovic M, Balvers JM. A fast method for the generation of boundary conditions for thermal autoclave simulation. Composites: Part A 2016;88:216–25. doi: 10.1016/j.compositesa.2016.05.036.
[9] Brauner L. Analysis of process-induced distortions and residual stresses of composite structures (Doctoral dissertation). Berlin, Germany: Logos Verlag, 2013.
[10] Xie G, Liu J, Zhang W, Lovenzini G, Biserni C. Simulation and improvement of temperature distributions of a framed mold during autoclave composite process. J Eng Thermophys 2013;22:43–61. doi: 10.1134/S1810232813010062.
[11] Hudek M. Examination of heat transfer during autoclave processing of polymer composites (Master's thesis, University of Manitoba, Canada), 2001. Available from http://mspace.lib.umanitoba.ca/xmlui/bitstream/handle/1993/18877/Hudek_Examination_of.pdf?sequence=1&isAllowed=y.
[12] Mazarro A, Garcia A. Tooling optimization for composite material parts. Paper presented at the European Congress on Composite Materials 17, Munich, Germany, 2016.
[13] Telikicherla MK, Altan MC, Lai FC. Autoclave curing of thermosetting composites: process modeling for the cure assembly. Int Comm Heat Mass Trans 1994;21(6):785–97. doi: 10.1016/0735-1933(94)90032-9.
[14] https://www.ainonline.com/aviation-news/aerospace/2014-02-10/boeing-and-airbus-partnerships-proliferate-asia.
[15] https://www.wisoven.com/products/batch-ovens/composite-curing-ovens.
[16] Dumont F, Fröhlingsdorf W, Weimer C. Virtual autoclave implementation for improved composite part quality and productivity. CEAS Aeronaut J 2013;4:277–89. https://doi.org/10.1007/s13272-013-0072-1.
[17] http://bloga350.blogspot.com/2015/11/32-meter-invar-tools-for-a350-wing.html.
[18] https://www.tencatecomposites.com/media/2eb7bb74-ca13-482c-9269-75c9964fe771/JrEIpw/TenCate%20Advanced%20Composites/Documents/Product%20and%20application%20one%20pagers/TenCate_AmberTool_Tooling_Backing_Structures.pdf.

[19] Baran, et al. A review on the mechanical modeling of composite manufacturing processes. Arch Comput Methods Eng 2017;24:365–95.

[20] http://pilotjohnsrv9.blogspot.com/2017/05/harvesting-clecos-for-wings.html.

[21] https://www.capacitec.com/content/editor/file/GapmanGen3_Aerospace_Applications_Artilce.pdf.

[22] Sarh B. Wing structural assembly methodology. SAE Techni Paper. 1998.

[23] https://blog.boomsupersonic.com/xb-1-s-wings-are-officially-closed-out-c9b53591252e.

[24] Drouot A. Measurement assisted assembly for high accuracy aerospace manufacturing. 16th IFAC Symposium on Information Control Problems in Manufacturing, 2018.

[25] Erdem I. Development of automated flexible tooling as enabler in wing box assembly. 6th CIRP Conference on Assembly Technologies and Systems (CATS), 2016.

Chapter 5

Manufacturing and quality

Chapter outline

This chapter will go into detail on aspects related to manufacturing and quality of composite materials that need to be understood in order to make a better informed decision about ways to design and manufacture tooling for those parts. The life cycle of a tool is complicated and involves many different phases each with its own complexities and nuances. Fig. 5.1 shows an outline of a tool life cycle and associated details. The manufacturing and quality phase involve many steps and are crucial in the success of the process and in order to achieve a final tool that adheres to all requirements and can support the tool purpose.

Depending on the material used to fabricate the tool, the manufacturing process can change quite dramatically including the level of inspection that needs to be done in order to accept the tool. And depending on what process is used to build the composite part (hand-layup, resin transfer molding (RTM), etc.) the tool will need to be designed in such a way to support that process. If the fabrication process was not well understood by the tool design engineer, this can cause the tool to be very complicated in nature and not adhere to the requirements needed. This chapter will serve as an introduction to several different types of composite manufacturing processes and provide some example tool designs. The reader is encouraged to use this content to support the design of

Tooling for Composite Aerospace Structures. http://dx.doi.org/10.1016/B978-0-12-819957-2.00005-5

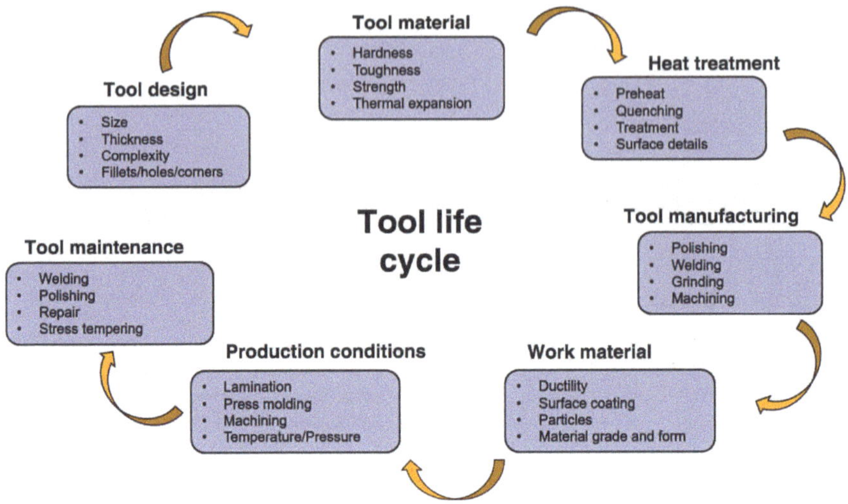

FIGURE 5.1 Tooling lifecycle.

future tools or the modification of existing tools as required. The list here is not a comprehensive manufacturing list rather a list of the most commonly used fabrication approaches in the aerospace industry mainly.

1 Composite manufacturing and tooling

Tooling is a foundation technology used in almost every major market and is often a routine part of the design and production process. With so many other factors to consider, it is easy to fall back on the default tooling methods and designs that have been used in the past. But for many organizations, moving to an alternate tooling material or using a new process for designing and developing tooling could help deliver more efficient designs saving both time and money. Across industries, more and more manufacturers are moving to materials that are lower in cost to fabricate with higher durability compared to metals in certain cases (composite, or "soft" tooling methods with materials such as polyurethane foam including additive manufacturing). While this may not be the right choice for all applications and tooling needs, there are numerous instances where using soft tooling either as the final tool or to aid in developing the final hard tool can help you reduce costs, iterate faster, or produce a more accurate tool.

Tooling is a critical part of the manufacturing process when it comes to parts made from composite materials. Using poor quality tooling will likely result in substandard components being constructed that are prone to malfunction, fail under stress, or unable to meet requirements or specifications. This can result in a large volume of parts being scraped or require repair.

The precision and characteristics of the tool have an impact on the quality of the finished part, its properties, the speed and accuracy with which the part can be repeatedly produced in high-volume runs. Using the correct process and materials to create those tools is critical to ensure properly functioning parts. To sum up, to create the best product, you must design the best tools, engineered to the highest quality, for the right job.

In manufacturing, tooling is the process of designing and engineering the tools necessary to produce the parts or components needed to develop the final product. It can include assembly tools such as jigs or fixtures; cutting tools such as milling and grinding machines, and welding and inspection fixtures. There are several methods for tool development and numerous materials, ranging from composites to hard metals that can be used to make these tools.

As production rates for composite parts are expected to expand rapidly in the next several decades in order to achieve the demands of the market, the quality target must be zero defects, zero rework and repair, and zero scrap. This is easier said than done. In order to achieve this goal a systematic change will need to occur in the process of how manufacturing occurs including the connection with design and assembly. The entire lifecycle of the product will need to be considered early on in order to try and streamline the procedures and define any drawbacks that might exist in the process. As a first step we will need to identify the major factors relating to variability of composite material products. The sources of variability have been broken down into materials, processing, and postmolding processes, which are summarized in Table 5.1.

Some of these variabilities will interact with the tooling and if the tooling was not appropriately manufactured that will only cause additional issues to the part with the possibility of scrapping it.

In spite of the increased market demand and opportunities for advanced composites, significant challenges exist in the United States for replacing traditional engineering metal alloys with composites. In a recent survey, respondents were asked to rank the top challenges when dealing with composites and the need for lower cost tooling came among the top 8 issues as shown in Fig. 5.2. This will be a long-term obstacle to the progress of composite manufacturing and enabling knowledge sharing across the industry will allow the removal of that barrier and enhance the understanding of the tooling details which this book aspires to do.

With many of the tooling aspects discussed thus far we will go into detail on composite manufacturing in this chapter. Without the understanding of the manufacturing of composites it will be very difficult to design and operate tooling specifically for lamination. The best engineers are certainly those that have a broad understanding of both manufacturing and tooling as they can make the right decision when time comes. We will focus on many different types of manufacturing including, hand layup, compression molding, RTM, etc., and show the different advantages and disadvantages of each approach and tooling methodologies for each. Before we go there we would like to introduce the reader

TABLE 5.1 Variability related to composite material fabrication.

Materials	Processing	Post-molding processes
• Degree of cure in the resin • Locking angle for cloth drape • Ease of wrinkle formation both in-plane and out of plane • Surface porosity prepreg • Surface smoothness prepreg	• Order of layup • Operator and supervision skills • Layup aids and tools provided • Tooling preparation quality • Tool release issues • Changes in tooling type • Temperature variations • Type of bagging materials • Bagging methodologies • Vacuum level in the bag • Cure cycle variations • Bulk resin flows • Local resin flows • Interactions between tooling and reinforcement due to CTE mismatch • Temperature at demold	• Deflashing process • Demolding procedures and forces • Operator, inspector and supervision skills • Uncertainty of datum • Edge trimming processes • Transport/carriage methods • Jigging for metrology and other postmolding processes • Machining and hole drilling processes • Difficulties in interpreting NDT results • Surface prep for bonding • Adhesive mixing and application for bonding • Cure cycle for bonding • Preparation for painting/ finishing • Application of paint/ finishing coats • Mechanical assembly processes

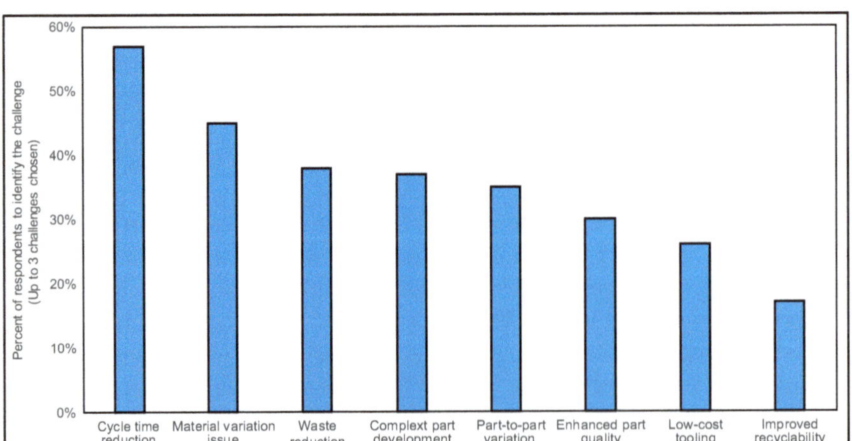

FIGURE 5.2 Challenges associated with the development of new composite manufacturing [1].

to details on composite manufacturing especially for those new to this subject via a case study. We will focus on the build process of a large wing skin that is typically used in many modern composite aircraft's.

2 Composite wing skin manufacturing: a case study

As stated in chapter 1, the use of composites have been widely adopted in many applications for almost four decades. Their superior properties such as light weight and high strength make them an attractive candidate for many products in different fields including aerospace, automotive and maritime applications.

The early use of composites in the aerospace industry was mainly for secondary nonstructural applications such as fairings and flight control surfaces in aerospace applications [2] as shown in Fig. 5.3. Given the tight regulatory nature of the aerospace industry, many certification and compliance details needed to be addressed prior to expanding the use of these materials across the airframe knowing that these regulations are constantly changing to ensure the safety of the general public which is their prime concern [3].

The certification requirements involve many different disciplines as shown in Fig. 5.4 and each one has its own requirement set. In order to have a smooth

FIGURE 5.3 Control surfaces and fairing made from composite materials [4-5].

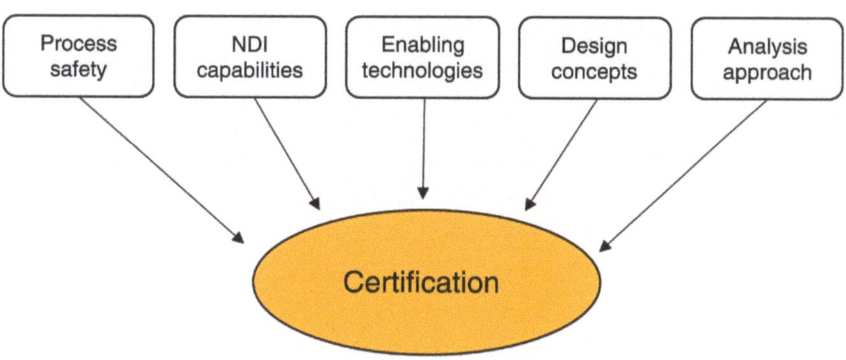

FIGURE 5.4 Different variables impacting the certification process in aerospace structures.

certification execution process, close collaboration is crucial among the different disciplines to ensure that all feedback is incorporated and prevent any details from being lost or forgotten.

These days, the use of composites in both commercial and military airframes exceeds 40% and are used for primary load path parts. That shift was possible due to the dedication of the industry to this innovation and the constant maturity in the analysis, material and process techniques as well as the fabrication approaches. With many successes came many setbacks and lessons learned [6-7] and the engineering community gained the confidence working with composites from those lessons.

Despite the technical challenges, the economical side had a large impact on embracing composites. The use of lighter weight materials compared to metallic meant that airlines and operators could save money on fuel cost during service. It was found that a saving of $46 can be achieved for every one pound saved in weight assuming $3.44 per gallon of jet fuel [8].

The main cost drivers when it comes to composites is the raw material, tooling, labor, manufacturing process, production volume, and equipment. Any improvements in those areas have the potential to drive the cost down. With advances occurring in those areas over the next few decades the cost of building parts from composites is believed to become more competitive with metallic materials which will make them an even larger competitor when it comes to future products [9].

As shown in Chapter 2, fiber polymer composites use many different types of materials including fiberglass, carbon/graphite, boron, kevlar, and other organic materials. They all exhibits properties of high specific strength but the most common is carbon fiber given its impressive properties across the board (mechanical, electrical, thermal). The fabrication process of composites has many different types including hand layup, automated tape layup, resin injection, compression molding, pultrusion, filament winding and many more [10]. The most commonly used is hand layup due to the level of flexibility it offers compared to other processes but can be more time consuming. As for composite joining there are other classifications which can be divided into three primary techniques summarized below:

Cocuring: The act of curing a composite laminate and simultaneously bonding it to some other uncured material, or to a core material such as honeycomb, or foam core. All resins and adhesives are cured during the same process.

Cobonding: The curing together of two or more elements, of which at least one is fully cured and at least one is uncured. Requires careful surface preparation of the previously-cured substrate. Additional adhesive may be required at the interface.

Secondary bonding: The joining together, by the process of adhesive bonding, two or more precured composite parts, where the only chemical or thermal reaction occurring is the curing of the adhesive itself. This approach requires careful preparation of each previously cured substrate at the bonding surfaces.

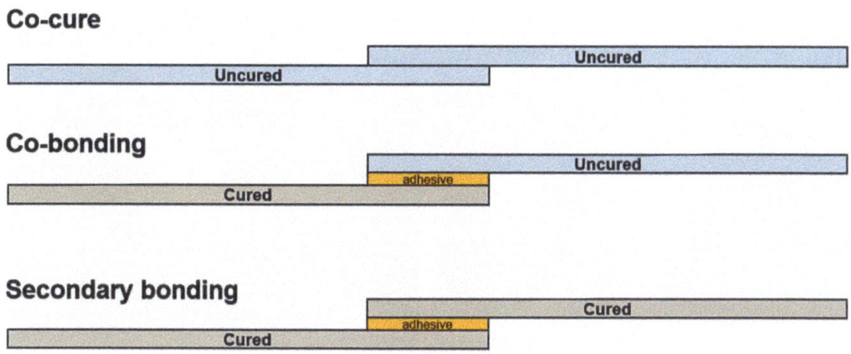

FIGURE 5.5 Classification of CFRP bonding techniques.

Usually requires well designed fixtures to align and clamp parts during processing. Re-heating previously cured substrates can be risky.

Fig. 5.5 shows a schematic representation of all three options. Details on each category is documented extensively in the literature [11] but as shown, the cocuring process requires only a single cure cycle and no use of fasteners to bond the mating parts which makes it an attractive option to reduce part count and ultimately cost. The assembly process including fastening is an extensive and complex process as shown in Fig. 5.6 and anything that can be utilized to minimize that is of great interest to the industry.

From the manufacturing standpoint, composites offer the unique ability compared to metallic materials of formability. Designs that would have never been thought possible using metallic materials are now possible. Cocuring has been the strength of composites and presents the best way of combining various parts without fasteners. Since the different parts are concurrently formed and bonded to each other, the assembly stresses are avoided. The reduction in number of parts results in reduced number of fasteners and reduced assembly cycle. Many different designs have been proposed in the literature when it comes to unitized composite parts [13-18] but very few have ever taken those concepts and built parts to evaluate the proposed designs. It has been shown that co-cured structures compared to cobonded structures have a more complex failure mode and are shown to be superior to bonded parts in many cases [19]. To ensure that the final co-cured part is well designed, a great deal of interaction between the design and fabrication teams is required. The final configuration of such parts should be decided based on its structural and fabrication feasibility [20].

When it comes to large primary structural parts (e.g., wings) there is always the need to add stiffening features to improve on their stability and strength as shown in Fig. 5.7. There are many different kinds of such stiffeners that have been widely used in the industry and each type behaves in a unique manner. Blade stiffeners are one type that have a great deal of use and many have evaluated their performance in detail [22] but the fabrication aspect is considered to be more

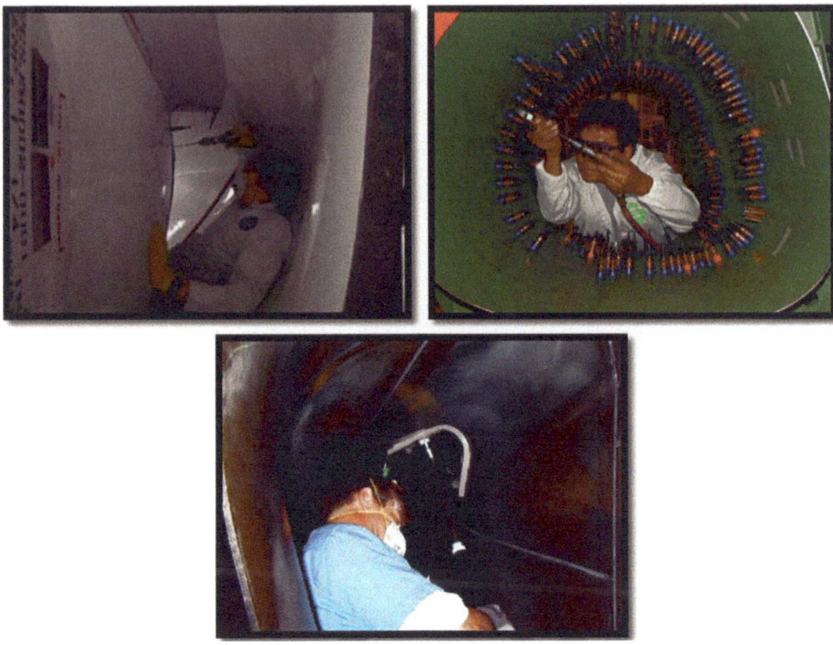

FIGURE 5.6 Complexity involved in the assembly process during manufacturing [12].

FIGURE 5.7 Wing skin with stiffeners [21].

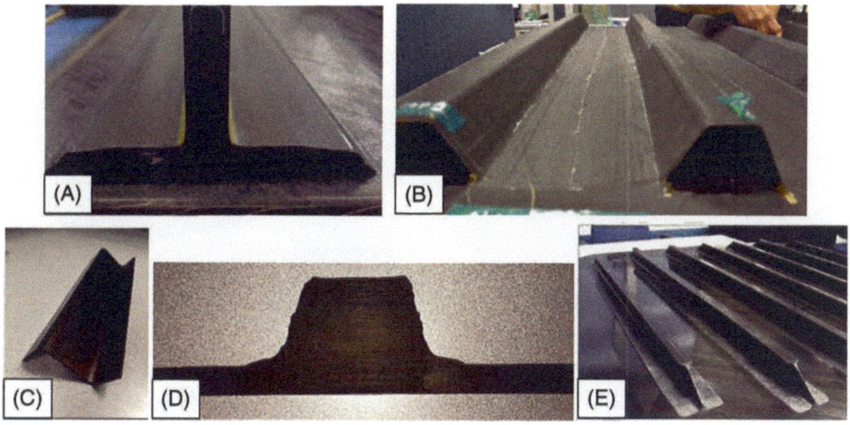

FIGURE 5.8 Different stiffener types (A) blade (B) hat (C) Z (D) plank (E) I.

complex compared to other options. Hat-shaped stiffeners are another widely considered stiffening feature typically used in aerospace parts. They are generally easier to build compared to blades including I shaped stiffeners. The cocure of those stiffeners to the mating structure is also shown to be less susceptible to structural delamination compared to others [23]. Therefore, for the build process of the full scale composite wing that we will discuss next will use a hat stiffener as our stiffening approach. Several other stiffening designs is shown in Fig. 5.8.

But what are the main inhibitors from using the cocured process more widely across the industry? The reason goes back to limitations such as the added difficulty in inspecting the parts for porosity and wrinkles and the complexity in the tooling approach among many others.

In order to overcome such limitations additional evaluations are required to investigate cocured processes especially when it comes to thick structures with clear documentation and recommendations across the industry. This applies to all other concepts that have large potential but lack of clear process documentation.

The fabrication process and tooling approach are proportionally related to how thick and large the part is and when it comes to wing skins specifically there are two primary tooling methods known as outer mold line (OML) and internal mold line (IML) approaches as shown in Fig. 5.9. Each approach has its own pros and cons that are summarized in Table 5.2.

Details on the manufacturing of the wing skin will be discussed next. For this study we will be using an OML approach given the simplicity of this method from the tooling side and the advantages it offers.

2.1 Wing skin fabrication

Prior to building any type of full-scale demo it is recommended to work on smaller-scale demos in order to perfect some of the details on lower cost parts

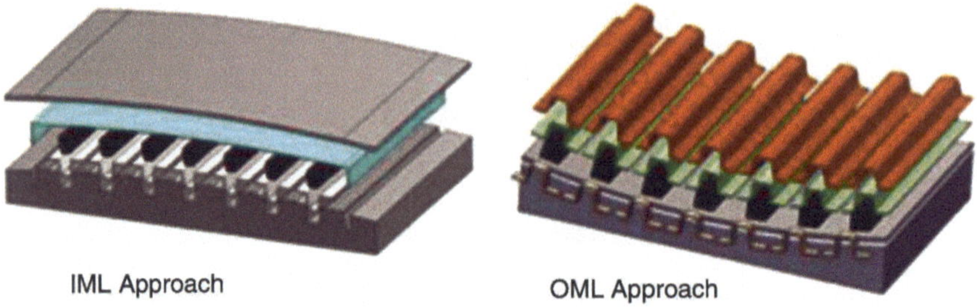

FIGURE 5.9 Tooling approach used for composite wing skins [24].

TABLE 5.2 Pros and cons of the different tooling approaches used for the build of stiffened composite structures.

Tool approach	Pros	Cons
OML Controlled	• Less complex • Less expensive • Can start tooling design as soon as the OML of the aircraft is established • More forgiving of design changes	• More labor • More risk for locating and maintaining locations of stiffeners • More difficult to bag
IML Controlled	• Less labor • Less risk for locating and maintaining locations of stiffeners • Simple to bag	• More complex • More expensive • Less forgiving of design change • Ability to adequately control the aerodynamic surface

and then take all lessons learned to apply to the larger scale. In this work, the team took a building block approach [25] in order to define specific details that will aid in the development process in order to minimize the risk during the full-scale build and aid in the development process. Although this section focuses on the full-scale demo it is helpful to briefly provide details on what was done on each building block level. For the coupon level, a complete test program was developed in order to finalize the allowable data set required for the material system used. On the element level, several flat panels were built in order to understand the level of effort required to build hat stiffened panels using the OML tooling approach in addition to identifying the complexity of each step during the fabrication process. Details on how to manufacture the tow filler were also established which is considered a critical detail when it comes to hat stiffened structures. Moreover, several test specimens were built to quantify the structural integrity of the hat stiffened structure. A schematic representation of

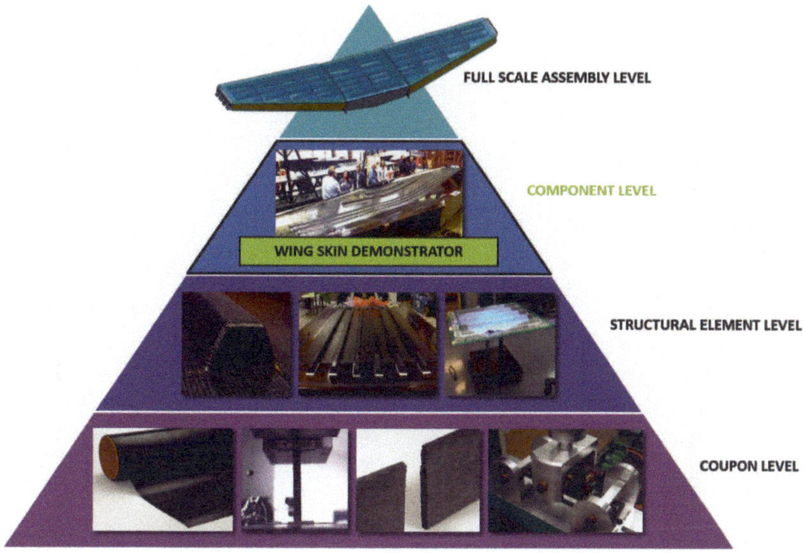

FIGURE 5.10 Manufacturing building block approach established for the program.

the building block details conducted throughout this study is shown in Fig. 5.10. Note that the full scale demo is part of the component level test.

2.2 Design details

A high-level overview of the size of the wing skin demo is shown on Fig. 5.11. In order to understand the different variables that will impact the quality of the part build we incorporated as many of those variables from the actual wing design into the demo wing skin design. Those variables can be summarized as follows:

Hat-stiffener size: two primary sizes were used in the demo. The first was approximately one inch tall and another that was 2 inches tall and both incorporate similar stacking sequence to eliminate that as a variable. The level of complexity in fabricating each type was of interest to the team in order to provide feedback as part of the design for manufacturing (DFM) feedback.

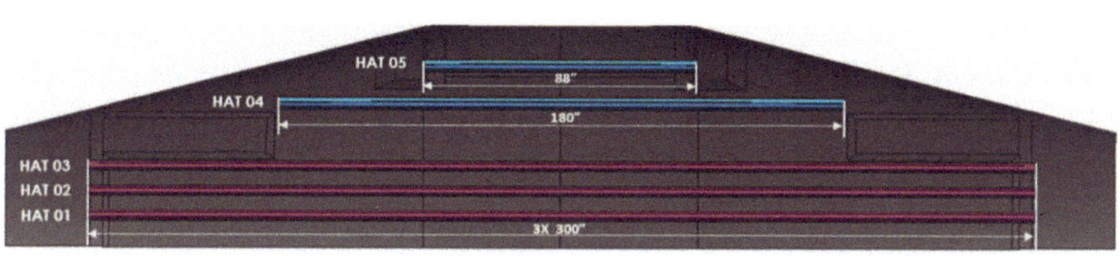

FIGURE 5.11 Wing skin demo design outline.

TABLE 5.3A Mandrel types used on the wing skin demo.

<table>
<tr><th>Mandrel type 1</th><th>Mandrel type 2</th></tr>
<tr><td>Compression molded mandrel
• Silicone (350°F rated)
• FEP coating (from manufacturer)
• 25-ft (qty 3)
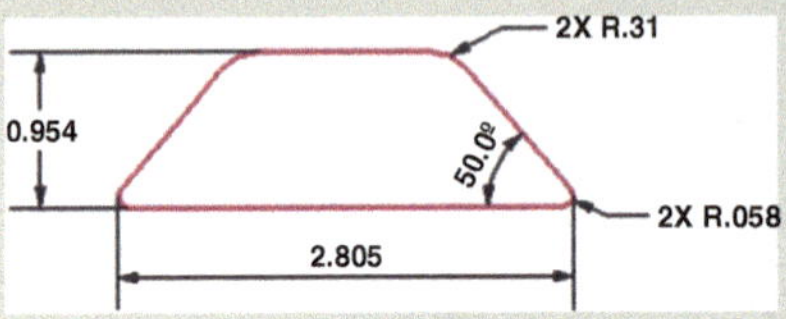
</td><td>Extruded mandrel
• Silicone (250°F rated)
• Wrap with Teflon tape (done in-house)
• 15-ft (qty 1), 7.3-ft (qty 1)
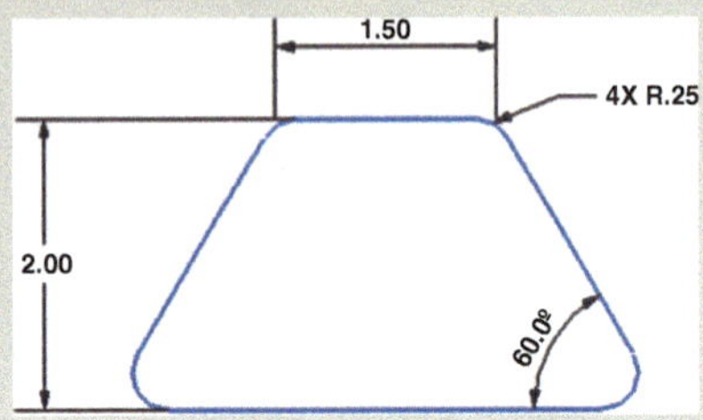
</td></tr>
</table>

TABLE 5.3B Hat stringer sizes used in the wing skin demo.

<table>
<tr><th>Proposed mandrel type</th></tr>
<tr><td>Compression molded mandrel with hole
• Silicone (350°F rated)
• FEP coating (from manufacturer)
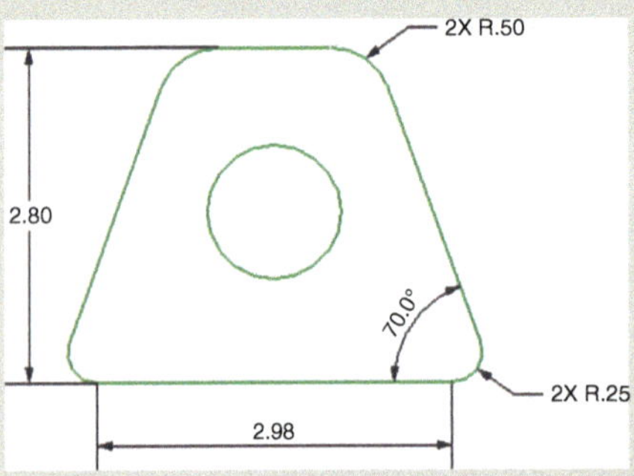
</td></tr>
</table>

Hat mandrel type: two primary types of mandrels were used for the build. For the smaller size, compression molded mandrels were used. Those are known to exhibit less expansion under temperature which is preferred in order to minimize the part distortion during cure. The other was an extruded mandrel and those have larger expansion but are known to be less expensive and have a quicker procurement time. Summary of both can be found in Table 5.3A.

Laminate thickness: since the actual design will incorporate many different laminate build-ups (BUPs) and thickness across the part we decided to include some of those details in the demo to evaluate the performance and identify

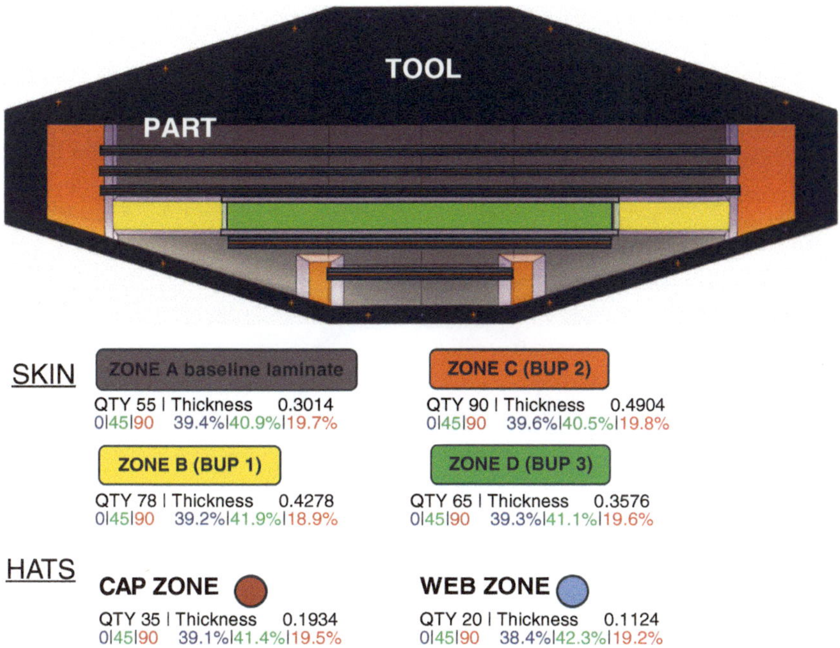

FIGURE 5.12 Skin and hat stacking sequence in the different regions of the wing demo.

shortfalls of any specific detail that needed attention. The skin thickness ranges from 0.5 inches to 0.3 inches which depends on the level of ply BUPs used in a specific region as shown in Fig. 5.12. A unique stacking sequence was used for the hats and were consistent among all 5 hat stiffeners.

Tooling material: in this study we used a composite tool that included a curvature that is more extreme compared to the actual design to ensure we capture all the complexities. The tooling material was LTM45 and the LM was scanned prior to any lamination to ensure it is within the engineering tolerances of +-0.015 inch across the part.

Sacrificial plies: another variable that will be evaluated in this demo is the use of sacrificial plies for the purposes of machining post cure. In order to ensure that the part is within the engineering tolerances post cure, the part is machined to fit the CAD model. The location where those sacrificial plies were included is shown in Fig. 5.13. Those areas typically correspond to joints where spar and rib interface exist.

In an ideal world there will be no need for sacrificial plies as the part should be fabricated per the design requirements but given the thermal gradients during cure, the unsymmetric and unbalanced behavior of the laminate, and the geometric differences across the part, those factors all lead to the creation of residual stresses within the part impacting the part warpage. In addition, human error and other material inconsistencies cause the part to lie outside engineering

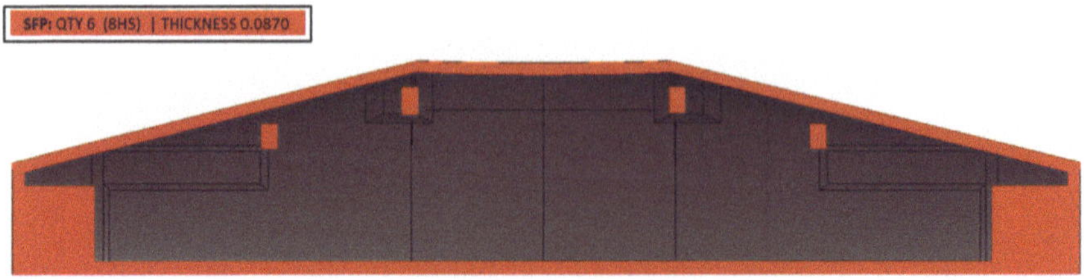

FIGURE 5.13 Sacrificial ply locations used across the wing skin demo.

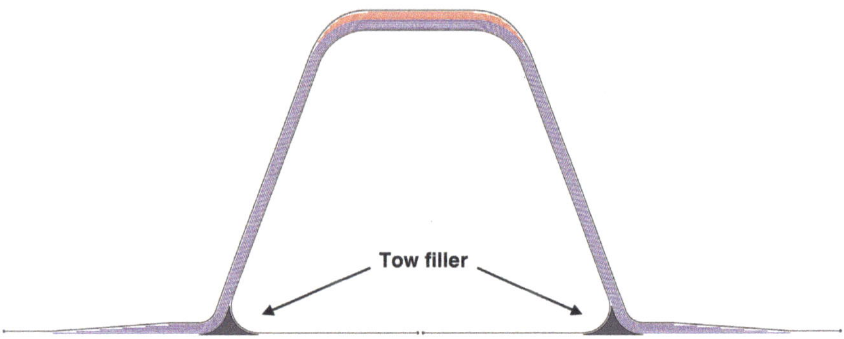

FIGURE 5.14 Tow filler used during the fabrication of composite hat stringers.

and design boundaries adding to the need for sacrificial plies. If one was able to prevent those factors we can envision a time where the use if sacrificial plies is eliminated.

Tow filler: the tow filler is used to fill the gap that is created when laminating the hat stringers as shown in Fig. 5.14. It is well known that the filler is considered a critical structural integrity of that part [26-28]. Hence, additional focus needs to be given to the fabrication process of this feature in order to reduce any anomalies that can impact its quality. We used two different materials for the evaluation, adhesive rolls made from FM309-2 and unidirectional Cycom 5320-1 with T650 fibers using a proprietary fabrication method. Both types will be evaluated as part of this demo.

2.3 Lamination process

The fabrication process utilized hand-layup as the primary method of lamination since it was considered our baseline process for building the part during production. For a more efficient process the use of AFP might be a better option especially during production and if rate is an important variable to consider. We will be discussing this later in the chapter.

These types of studies allow the development team to capture any critical details that need to be adjusted in the process specification prior to building the

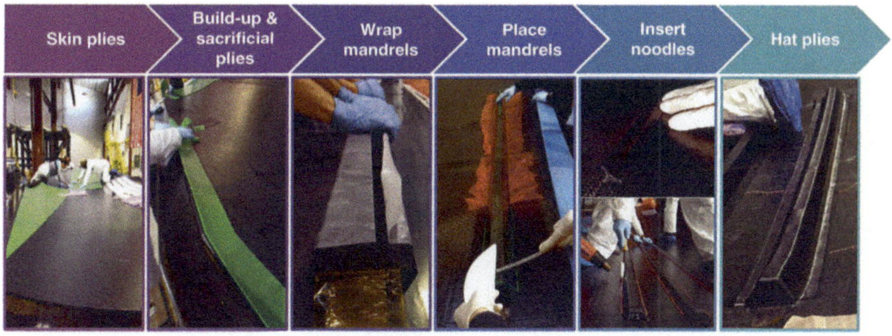

FIGURE 5.15 Overview of the lamination process used for the wing skin demo.

production parts. A high-level overview of the lamination process is shown in Fig. 5.15.

The skin included 55 plies overall with the addition of 35 plies in the build-up regions. The thickness was 0.3 inches with a max thickness near build up regions of 0.5 inch. The overall stacking sequence was a 40/40/20 laminate design utilizing all unidirectional material form. Plain weave was used as cover plies on the OML and IML surface of the skin and 8HS material form for the sacrificial plies.

FiberSim [29] was used to perform the ply splicing for this effort. The total number of individual plies was more than 1300 which is considered to be a very large number by all standards. Therefore, the handling of that many ply quantities as well as ensuring that the material out-time is not exceeded (20 days for handling and 30 days while on tool under vacuum) was a critical requirement. During the layup it was noted that the ±45°ply orientation which had a size 42 inch wide x 98 inch long was manageable to handle with minimal issues overall. The same applied to the 90°ply orientation of size 42 inch wide x 70 inch long. On the other hand, the 0°ply orientation size 42 inch wide x 200 inch long caused difficultly during layup in the form of wrinkles. Fig. 5.16 shows the lamination process of all three ply orientations used.

The hat stringers had a total of 35 plies with thickness of 0.2 inch (20 web plies and 15 cap plies). We used rubber mandrels from Rubbercraft, LLC, as the preferred product for this demo. The lamination occurred one ply at a time using hand layup. A similar 40/40/20 laminate design was used with all unidirectional material form for the exception of PW cover plies on mandrel and IML surface. In order to ease the post cure inspection specifically the blue light scanning to measure the movement of the hats pre and post cure a pre-impregnated peel ply (P15448) was used on the IML side. The hat final shape is shown in Fig. 5.17.

Some of the challenges that were observed during lamination can be summarized below:

- Compound curvature induced wrinkles in 0°skin plies as shown in Fig. 5.18

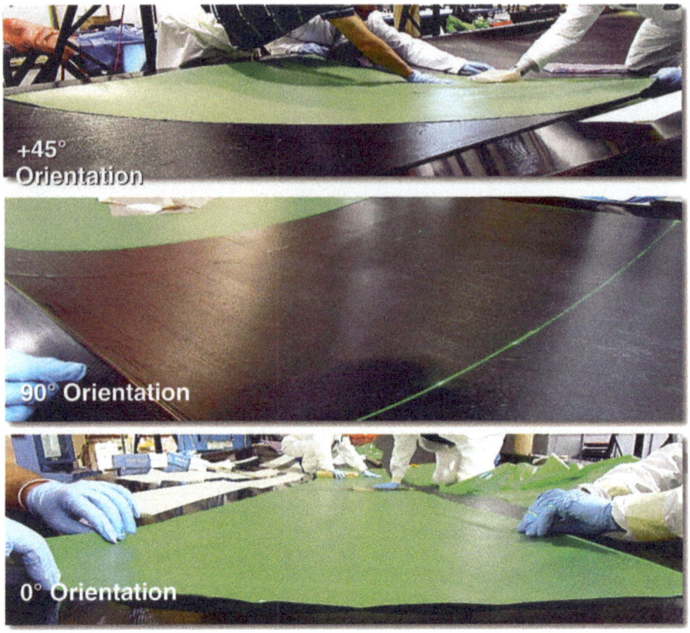

FIGURE 5.16 Lamination process of the three main ply orientations.

FIGURE 5.17 Hat lamination final product.

- Demo utilized maximum material width of 43 inch which was difficult to handle
- Working out wrinkles was difficult and time consuming
- Inability of FiberSIM to predict the wrinkles in the part using the set threshold values.

Improvement opportunities:

- Adjust FiberSIM properties for unidirectional material
- Splice material to smaller widths
- Provide manufacturing training to increase familiarity with FiberSIM generated ply kits on the floor

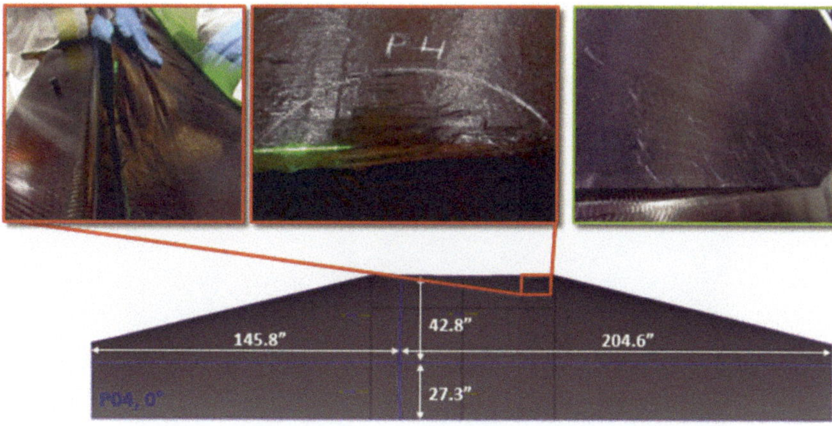

FIGURE 5.18 Fabrication difficulties observed during the fabrication process of the wing demo.

2.4 Mandrel placement process

Once the skin was fully laminated, the hat mandrels with the initial wrap ply had to be transferred on in the appropriate location per design requirements. For this demo the mandrel was placed on an FEP layer and wood support for transport. Laser projections were made on the skin and the tool was transferred near those laser points as shown in Fig. 5.19. Footprint projection were used for initial placement and centerline projection were used for final quality assurance (QA) verification. Mandrel centerline was hand-marked onto a wrap ply after poly removal. The FEP was then carefully pulled out from under the mandrel and the mandrels carefully moved back to center line. Mandrels were taped in place to reduce mandrel movement during debulk which was done after all the hats were placed in position. It was noted that the mandrels were slightly raised above the part surface at the BUP ramp locations prior to debulk. An approximate gap of 0.125 inch was seen but after debulk the gap closed with no issues as shown in Fig. 5.20.

2.5 Tow filler placement process

After the five hats were placed we finalized the lamination process by dropping the tow fillers. As stated earlier, two material systems were use, pre-formed UNI tows on HAT 04, HAT 05 which were five feet in length as shown in Fig. 5.21. Splices were scarf jointed at a 45° angle as shown in Fig. 5.22. The tow filler conformed to the tool curvature with no heat required. Teflon tools used to secure tows into the mandrel radius. Hand rolled adhesive film were used on HAT 01, HAT 02, HAT 03. Material strips were folded into 1/4" strips on a heated table and then rolled to final shape. HAT 01 had a 1.0 inch wide strip while HAT 02, HAT 03 had a 1.5 inch wide strip. Tows were heated with a heat gun and formed into the mandrel radius using Teflon roller.

FIGURE 5.19 Mandrel placement during the wing skin demo fabrication utilizing laser projection.

FIGURE 5.20 Gaps generated after placement.

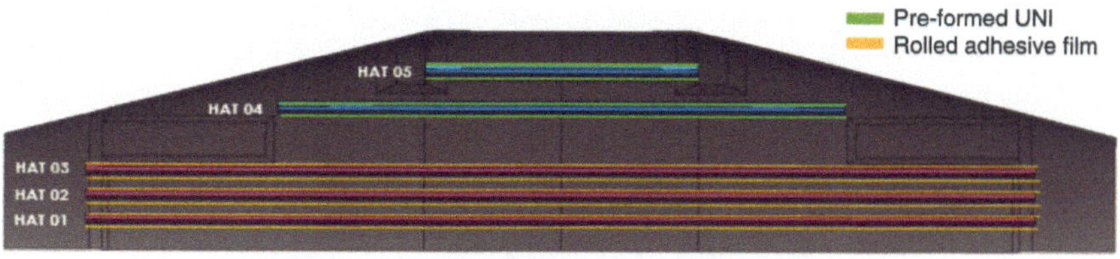

FIGURE 5.21 Tow filler types used on the wing demo.

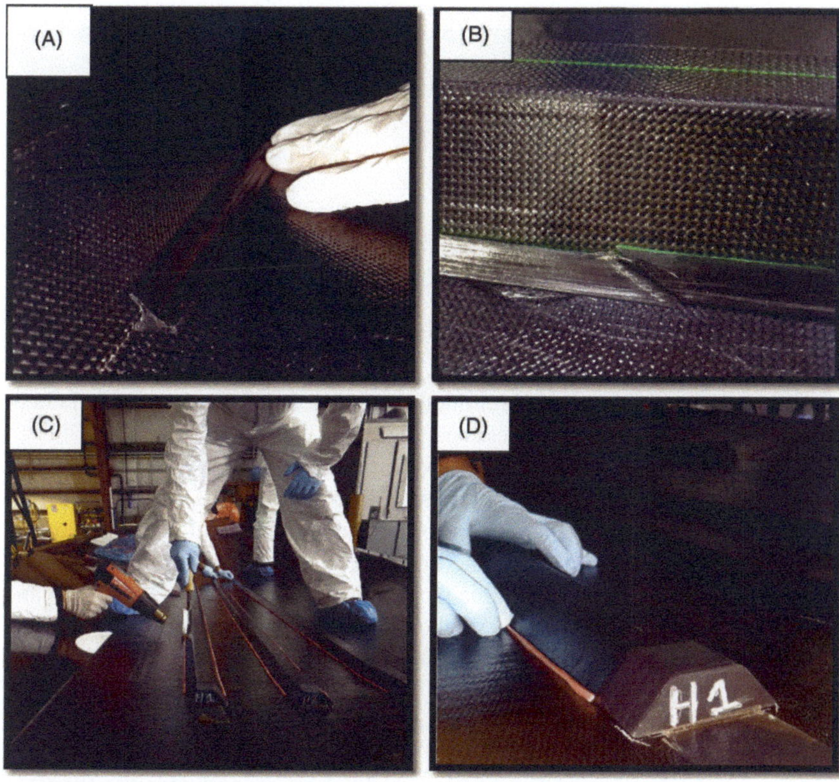

FIGURE 5.22 Details on the fabrication process of the tow fillers used in the demo: (A/B) UNI material (C/D) adhesive material.

2.6 Bagging and cure

After the completion of the lamination process comes the need to perform the appropriate bagging to prevent any leakage during cure. Bag leakage is a well-known nonconformance that typically occurs during the manufacturing process and might cause the part to get scrapped [30] which begs the need for investigation prior to building parts in production. Given the sheer size of the full scale

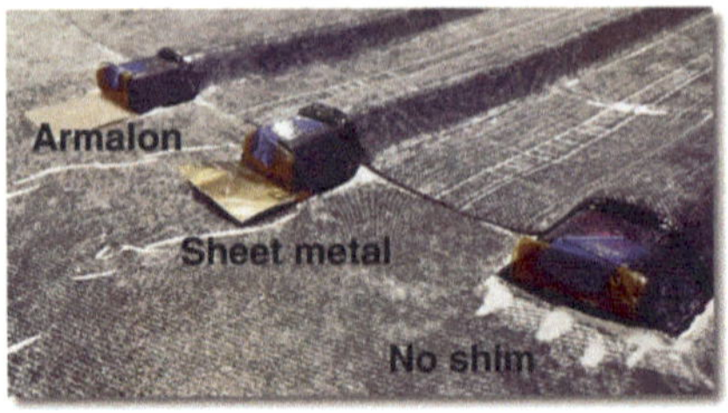

FIGURE 5.23 Shim material types used for the wing skin demo.

FIGURE 5.24 Thermocouple location across the part.

demo, unique challenges were identified that needed to be addresses and documented for inclusion in the process specification.

In addition, since this is a fully cocure process and given the thickness changes across the part an uncontrolled exotherm could cause the part to cure from the inside out, causing severe process-induced residual stresses. To prevent that, detailed evaluation of the cure cycle and temperature distribution should be captured to understand the thermal gradients observed in the part and provide a recommendation on the final cure prior to production build.

In order to enable the extraction of the mandrels post cure the mandrel ends were shimmed as shown in Fig. 5.23. Three different approaches were used: HAT 01 had no shim while HAT 02 and HAT 04 had sheet metal shim with teflon tape and finally HAT 03 and HAT 05 had an Armalon shim.

A total of 10 thermocouples at various locations were embedded as shown in Fig. 5.24 in order to ensure we capture the entire thermal gradient of the part appropriately. The part was initially covered with one FEP layer tailored over hats followed by one layer of N4 breather over hats and two layers of N4 breather in nonhat regions. As for bagging material, the choice was a nylon bag (Solvay HS8171-666 V-SHEET). Pleats were added at noodle radii of each hat and wing kinks as shown in Fig. 5.25. Ten vacuum ports were utilized with five on each side approximately 6-8 feet apart. The part was then checked for any leakage using a threshold of no more than 1 inch-Hg drop in 5 minutes.

The material used in this study for fabricating the part is Cycom 5320-1 epoxy resin system with T650 fibers. Given the large thickness of the part we

FIGURE 5.25 Bagging pleats used around the part.

needed to ensure first that we have a cure cycle that will lead to a fully cured high-quality part. RAVEN software [31] currently has this material system fully characterized and we took advantage of that fact to evaluate many cure cycles and come up with an ideal case for the specific details pertaining to this demo. Fig. 5.26 shows a specific cure cycle and the corresponding degree of cure (DOC) which reaches an appropriate value which provided us with enough confidence to pursue this cure recipe for the demo part.

The actual cure cycle used on the part is shown in Fig. 5.27 including the readings from all the thermocouples. As noted, we did not observe any exotherm during the cure. The specification allowed for a temperature range of 350 + /-10F and as noted the cure cycle maintained within that profile

2.7 Postcure evaluation

In order to quantify the amount of movement the hat stiffeners observed after cure, a precure and postcure scan of the demo was taken using blue light technology and analyzed the level of displacement as shown in Fig. 5.28. It is extremely important to ensure that the movement does not exceed the values set by engineering as the parts may not be able to fit during assembly (e.g., rib onto the skin). The engineering tolerance set for this study was a nominal sealing gap between the rib and hat of 0.140 inch as shown in Fig. 5.29.

The comparison was done between the post cure part and the CAD model design which ideally should be within the set engineering tolerances. Fig. 5.30 shows that comparison, note that the scan was taken while the part was still

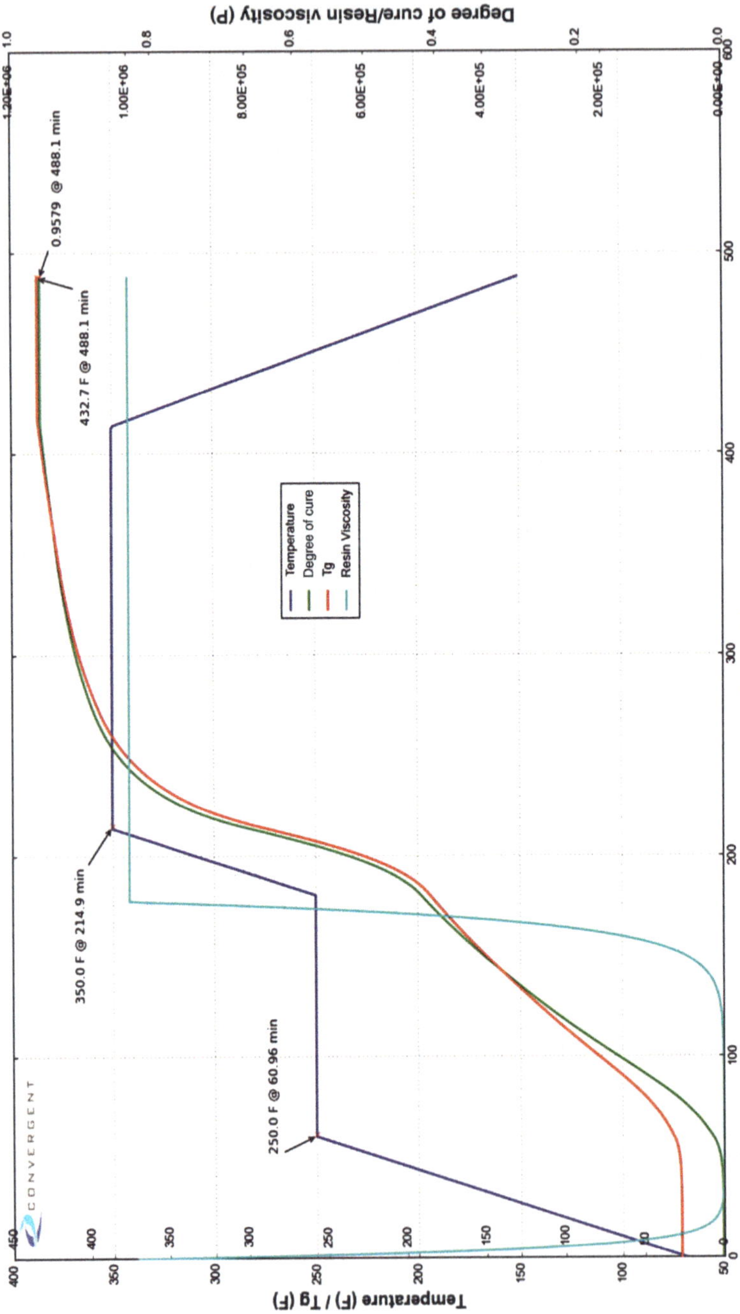

FIGURE 5.26 Cure cycle analysis for Cycom 5320-1.

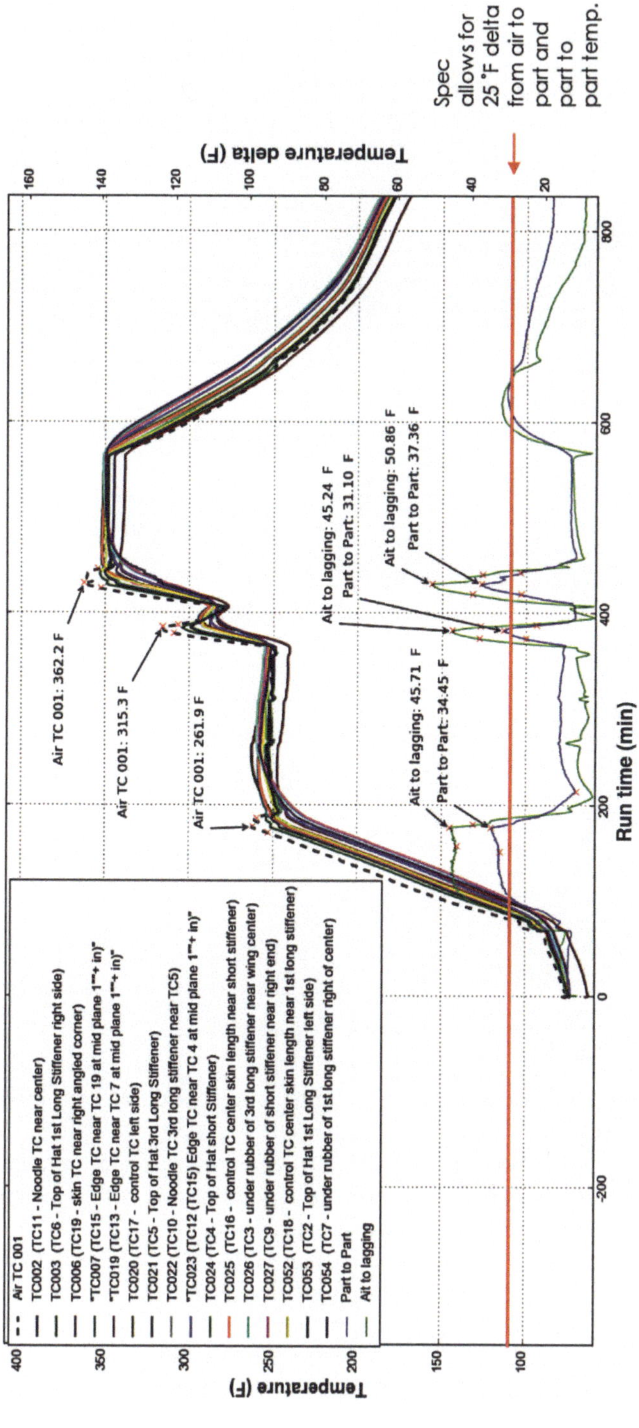

FIGURE 5.27 Actual cure cycle of the part and the thermocouple readings.

FIGURE 5.28 Blue light used to measure the part deformation precure and postcure.

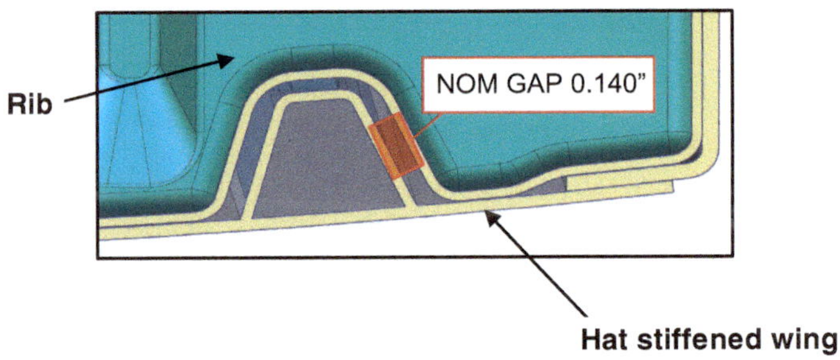

FIGURE 5.29 Nominal gap required between the mating parts in the wing structure.

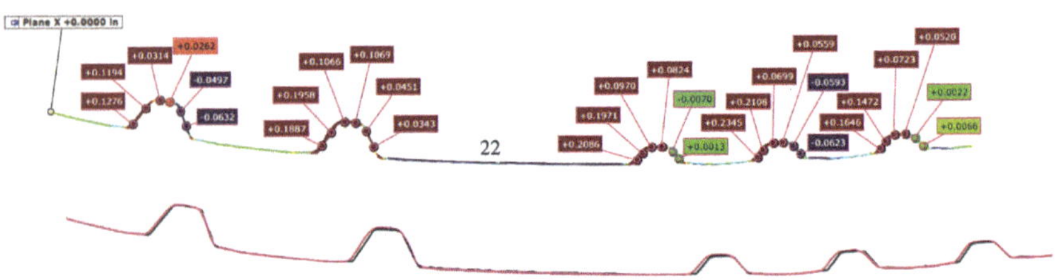

FIGURE 5.30 Post cure comparison to the CAD model.

on the tool and vacuum applied to eliminate any residual stress effects on the measurements.

The max profile deviations of the hat stiffeners was approximately +0.208 inches but for the majority of the part the deviation was within the engineering tolerances set. There are several potential reasons for the excessive movement in those areas. One of those can be attributed to the blue light scan data which has potential inaccuracies in the best fit process. Another is the deviations due to the initial placement of the hat stringer onto the skin. Recall that the hat stiffeners were placed based on a laser projection and that may have caused errors in placement. One remedy to this issue can be the additional use of tooling or shop aides required for mandrel positioning. An additional reason for the deviation can be attributed to the mandrel expansion. Depending on the type of mandrel, the expansion ranges up to 5% for the extruded mandrel on each web wall and up to 3.5% for the compression molded mandrel on each web wall. Fig. 5.31 shows a scan of a hat stiffener and comparison to the CAD surface postcure. As shown, the mandrel expansion impacts the final shape of the part across the perimeter. Finally, since the CAD model does not account for mandrel expansion we can expect some differences due to that when performing the comparison.

Another detail worth noting is the amount of bowing caused due to the mandrel expansion. As shown in Fig. 5.32 there is minimal bowing caused due to the expansion (0.03 inch) and this can be improved on by making some changes in the design of the mandrel namely including a hole in the mandrel as that will minimize the amount of expansion occurring as shown in Table 5.3B.

Fig. 5.33 shows the results of using the shim at the end of the mandrel termination. As shown, the plies under the metal shim behaved much smoother compared to the mandrel that did not include any shim. The waviness seen can cause premature delamination of that region under load causing structural damage.

Another part of the evaluation was the nondestructive inspection. A C-scan of specific critical areas in the part was done in order to identify the part quality and ensure it is free of defects. Given the lack of large scale equipment to scan the entire part and time associated with that the focus was only on areas of concern related to the design. Those areas included the hat terminations, large curvature regions, locations where wrinkles were seen during layup, build-up

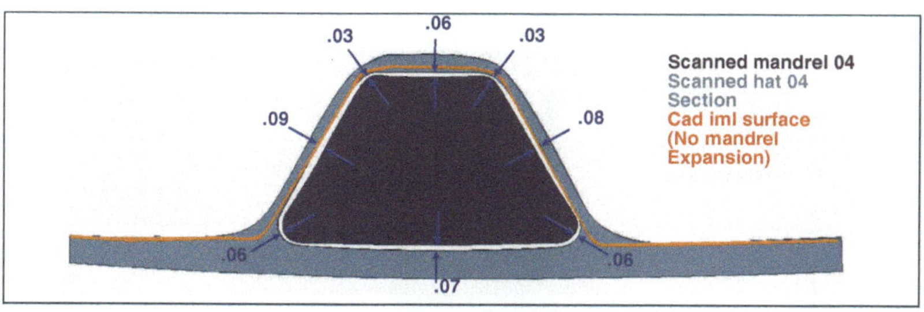

FIGURE 5.31 Mandrel expansion effect on the final part shape.

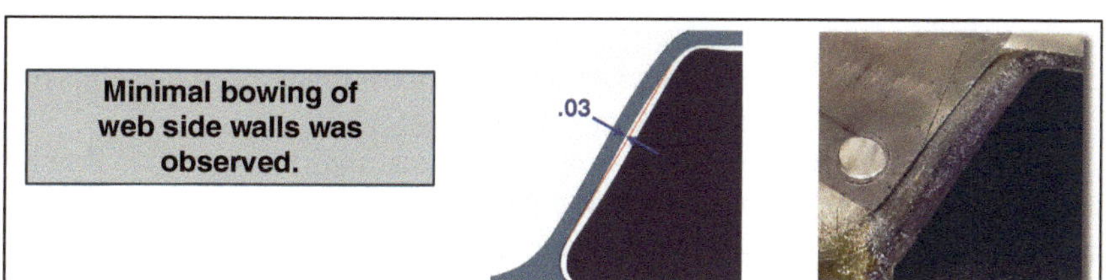

FIGURE 5.32 Bowing due to mandrel expansion.

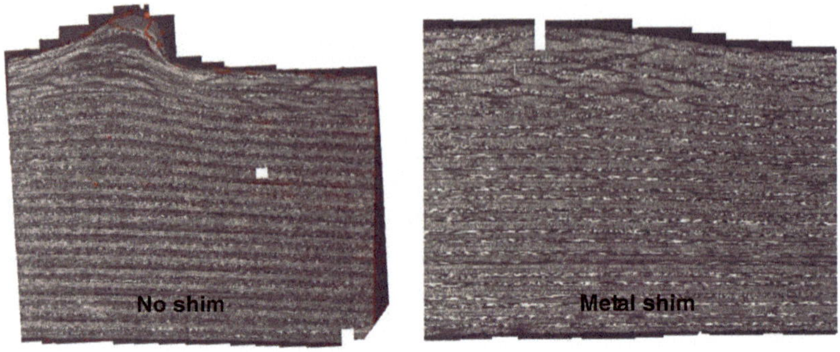

FIGURE 5.33 Hat termination behavior due to the use of shim material.

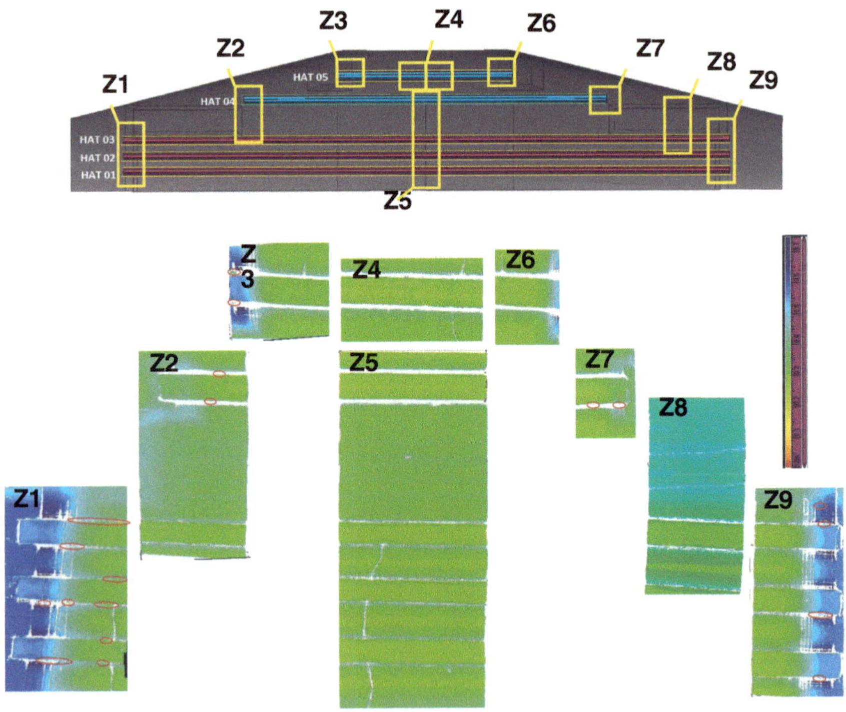

FIGURE 5.34 Nondestructive inspection results of the wing skin demo.

regions and ply drop terminations near hat web plies. Fig. 5.34 shows the areas where the scan was taken and the corresponding results.

It can be noted that zone 1 included inter-laminar indications within the skin over the tow area while zone 2, 7, and 9 had minimal inter-laminar indications. Zone 3 shows some indications near the hat termination. In order to correlate those indications to a level of damage or porosity destructive inspection was conducted. Fig. 5.35 shows a cross section cut near Zone 2 and the level of

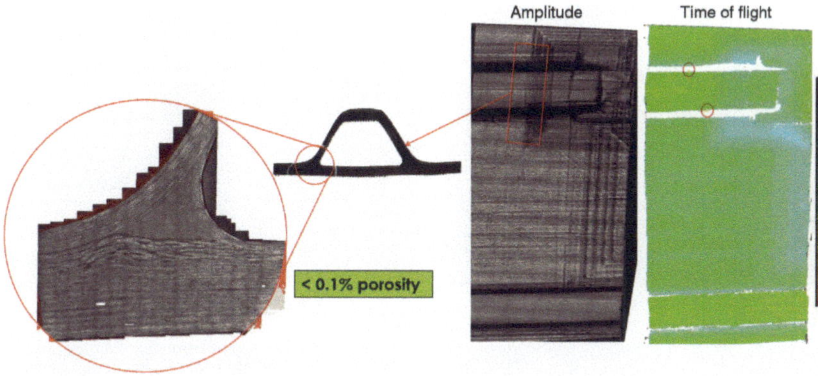

FIGURE 5.35 Micro-section cut of Zone 2 to correlate NDI with porosity.

porosity seen in that region. As shown, the porosity level was less than 0.1% which is much lower than the threshold specification limit of 2%.

Another portion worth evaluating was the behavior of the sacrificial plies as they are a large component of the actual part build. Recall that the sacrificial plies used 8HS material form. By evaluating the region where the sacrificial plies were included we notice that the one inch tall compression molded mandrels imprinted approximately 0.050 inch near the termination while the 2 inch tall extruded mandrels imprinted approximately 0.090 inch as shown in Fig. 5.36. Ply waviness was generated under the tow filler where the sacrificial plies existed. Comparing this location to areas where there were no sacrificial plies we notice that no waviness exist per Fig. 5.36 which suggests that the use of 8HS as a ply material sacrificial is not necessary the right option and it might be more effective to use unidirectional material instead to minimize any incompatibility in the laminate.

From this work it was shown that the use of an OML tooling technique to build a full scale wing skin is feasible and that is encouraging given that it's the preferred method when it comes to tooling since its much less complex compared to the IML approach.

With this detailed review on the pre and post fabrication analysis of this full-scale wing demo the reader shall now have appreciation on the complexities involved in composite material manufacturing. There are many other manufacturing techniques available and we will go into several well-known techniques and discuss their associated tooling approaches as well.

3 Resin transfer molding (RTM)

RTM is a process with a rigid closed mold where layers of dry fabrics known as preforms are inserted into the mold and resin is injected to generate the final part shape. It allows the part to be built as a unitized piece rather than building it using traditional autoclave methods such as hand layup. Fig. 5.37 shows an

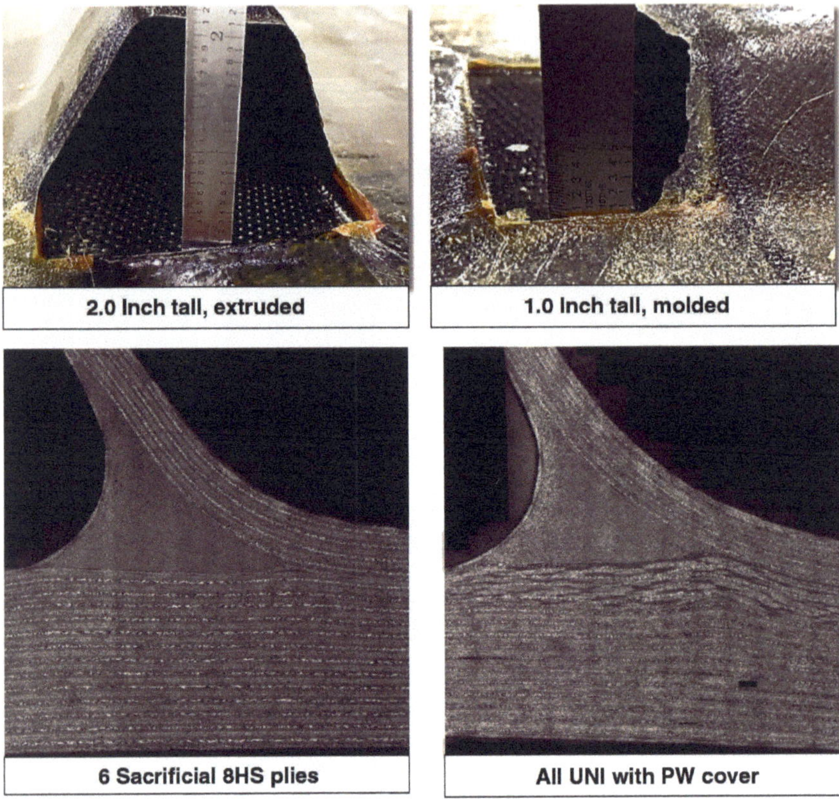

FIGURE 5.36 Impact of the sacrificial plies on part quality.

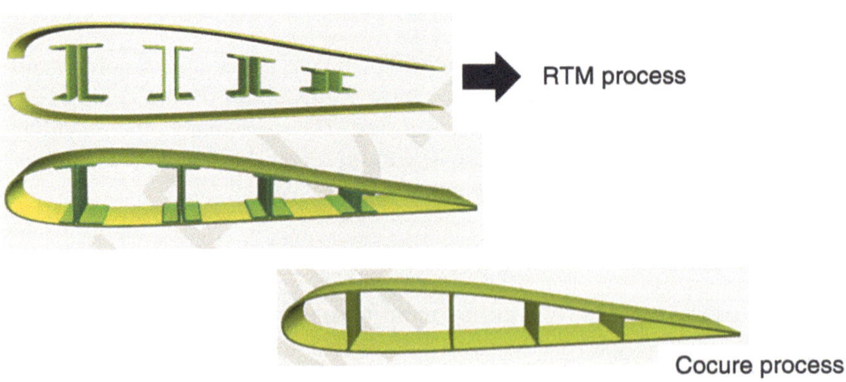

FIGURE 5.37 Comparison between traditional and RTM process [32].

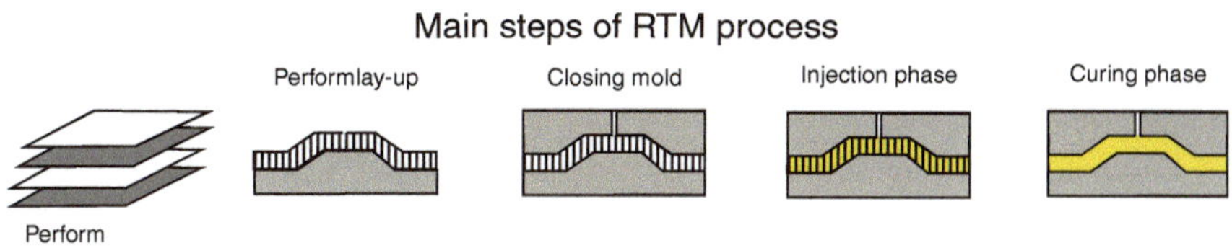

FIGURE 5.38 RTM process steps.

example of an aircraft winglet built using the traditional techniques that include several parts all assembled together and a net shape RTM process.

The main steps associated with RTM is shown in Fig. 5.38. First, the preform is draped in a half-mold, then the mold is closed and the preform compacted. After that, the resin is injected using a positive-gradient pressure through the gate points replacing the air entrapped within the preform and vacuum is applied at dedicated vents in order to favorite the air escape from the mold. When the resin reaches the vents, the gates are clamped and the preform is impregnated. After having the resin fully enclose the part enters the cure phase. Finally, the mold is opened and the part is removed.

Having matched closing molds allows for the compaction of the fiber reinforcement and allows the part to reach the desired thickness and fiber volume fraction. The compaction changes the microstructure and the dimensions of the preform, producing large deformations in some cases. The injection phase must ensure the complete impregnation of the preform or many anomalies may be present in the part including dry spot areas with missing adhesion between the layers, which makes the surface rough and irregular. These anomalies can cause issues in the top-level assembly, which are costly to resolve which is why close attention needs to be given to every step in the process.

3.1 Process parameters

The RTM process is governed by many different variables and parameters that are dependent on each other and their combination impact the process and the quality of the finished part. The most important parameters when it comes to the design are the pressure, temperature, viscosity, permeability, volume fraction, and filling time of the process. There are many other parameters that are independent of the previous parameters but need to be considered such as the angle of attack of the nozzle, the orientation of the fibers, the paths of flow and shear rates, the stratification and so on. The resin tends to flow more quickly in the fiber direction, thus the flow dynamic depends mainly on the type of fabric used and the part thickness. The thickness becomes a critical design constraint especially in the case of the inclusion of stiffening features and ribs. The injection velocity relies on the injection pressure and how the resin flows into the mold. The velocity also impacts the hydraulic pressure and the holding and closing forces of the mold. Consequently, the injection velocity defines the filling time, which should not be too short to ensure an adequate impregnation of the fibers and at the same time, the filling must be such as to avoid the risk of premature gelation of the resin.

The potential formation of air voids in the matrix, the appearance of surface anomalies, and the mechanical properties of the finished product has a strong correlation to the injection pressure. Another phenomenon in which this parameter is relevant, together with the viscosity, is the so-called "fiber wash," that is, the movement of the reinforcement inside the mold during the injection phase. In this case, the surface treatment of the fibers and especially the choice of the binder play a fundamental role. If the binder dissolves too quickly in contact with the resin, then fibers under the injection pressure can move freely. Temperature is an extremely important process parameter and it is related to the injection pressure and the viscosity of the resin. When the temperature increases, the filling time decreases and the working pressures are lower. When the temperature is low, the viscosity of the resin increases and it is necessary to increase the pressure to ensure the transfer of the resin itself.

The selection of the preform is another important aspect of the overall design that becomes very important for RTM process success. These preforms are prepared separately and constitutes the skeleton of the final product, greatly simplifying the molding operations and reducing the time and cost of processing. Example of a 3D preform is shown in Fig. 5.39. This allows for the building of net-shape 3D-complex structures. This possibility is given by the development of the preform technology.

FIGURE 5.39 Example of composite preform used in RTM [33].

The preforms are manufactured using textile like techniques. The choice of one method over the other depends on several factors: the processability, the feasibility of the geometry, the desired mechanical properties of the molded part, the cost of production and the performance required by the final application. The choice of the architecture of the fiber reinforcement depends on the required performance of the composite structure and the characteristics related to the process, such as permeability, compressibility and drape.

The complex fiber architectures can be obtained with the weaving method by interlacing and knitting the fibers along the three spatial directions; an example is shown in Fig. 5.40. Different bi-axial layers can be also stitched. The stitching method consists on darning the layers with fibers. These are automated techniques that realize complex shapes and 3D junctions in place of bolts and rivets. From the mechanical point of view, these methods can increase the crack resistance in composites by introducing fibers in the through thickness directions preventing cracks from growing. On the other hand, the stitching seams can produce local defects induced in the preform as a result of penetration of

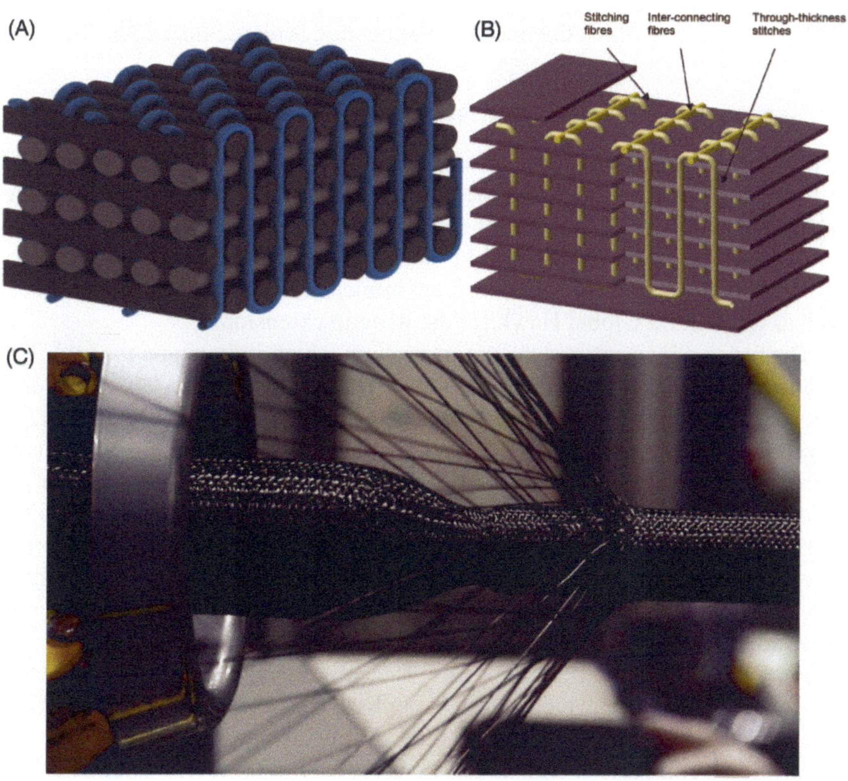

FIGURE 5.40 Preform weaving (A) 3D weaving [34] (B) stitching [35] (C) braiding [36].

the wire and the needle. Furthermore, the robotic system is very expensive and sometimes damage due to misalignment of fibers can occur.

For the RTM process thermosetting resins are the matrix used due to their low-viscosity during processing. There are many different types of thermosetting resin for RTM application and most of the process parameters (e.g., temperature, pressure, etc.) cannot be selected without considering the chemistry of the resin to be used. Factors to take in consideration for a RTM system can be divided into two broad categories: processing and performance. Initial viscosity and molding life are function of the temperature, and they determine the operational temperature range of a process. The molding time is a function of the rate at which the reaction occurs between the resin and the curing agent and the rate is directly proportional to the temperature. The viscosity depends on the chemical-physical characteristics of the matrix. Viscosity may change over time because of both temperature variations and as a consequence of chemical reactions that occur in the liquid state. The knowledge of the rheological behavior of the system is essential for a proper setting of process parameters. The values of the viscosity for the resin system needs to be adjusted to guarantee both the simultaneous removal of dissolved gases and moisture entrapped in the matrix, and the compaction of the fibers, before reaching the gel point. However, the viscosity of the resin must not be too high, especially in the case in which the fiber volume fraction is higher than 40%-50%.

4 Same qualified resin transfer molding (SQRTM)

This process was developed by Radius Engineering Inc. It is similar to RTM in that it uses a closed molding method but combines prepreg processing and liquid molding to produce true net-shape, highly unitized aerospace parts. The claim in this method is that SQRTM is designed to produce an autoclave-quality part without the autoclave and it has been implemented on some actual products such as the RQ-1B Global Hawk for the wingtip extension.

The process incorporates the use of vacuum which is drawn on the tool. The heat-up/ramp rate and cool-down follows prepreg fabrication specifications. Prepreg is tape-laid, drape-formed, or hand-laid, then debulked under vacuum, as per existing process specs. The part is transferred to matching tools made of invar, steel or aluminum. Tooling is clamped in a press and instrumentation attached. Once the tool is heated, small quantity of prepreg resin is injected into the tool to fill tool cavity around the edges of the part and does not impregnate the prepreg, only creates fluid pressure. The resin is intended to maintain a steady hydrostatic pressure within the mold. The pressure keeps volatiles and water vapor in solution to prevent void formation. The resin hydrostatic pressure maintained at 6-7 bars during cure. Because the higher thermal conductivity of the press and tool which permit faster heating and cool down, the SQRTM cure cycle can be as much as two hours shorter than an autoclave cycle. An example cure comparison with autoclave is shown in Fig. 5.41. It shows that the time reduction can be up to 2 hrs which is significant for production throughput.

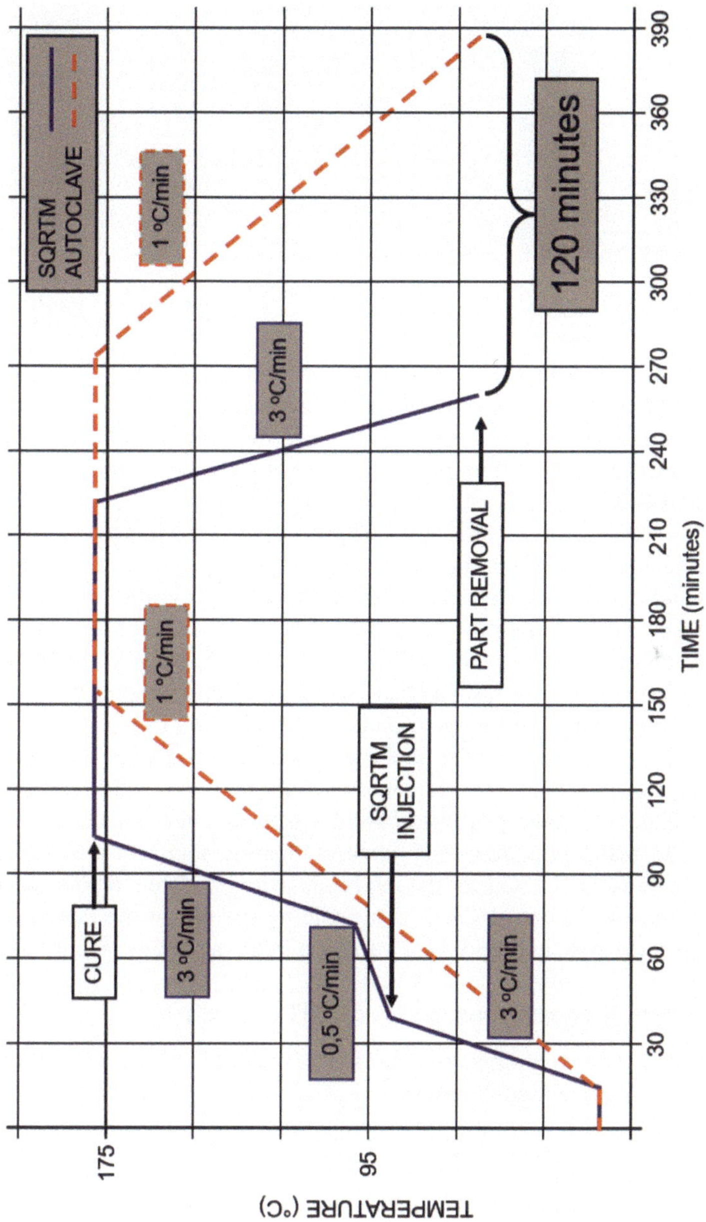

FIGURE 5.41 Cure cycle comparison with autoclave [37].

FIGURE 5.42 Lamination process on the closed mold tool [37].

An example of using this was an extremely complex, one-piece prototype helicopter cabin roof, produced under the Survivable Affordable Repairable Airframe Program (SARAP). We will go over the process for the helicopter cabin roof in order to provide the reader with content on the process and the tooling approach.

The first step involves the layup of the complex roof part as shown in Fig. 5.42 which involve the placement of a combination of debulked prepreg materials and dry preforms on the lower mold half. This is followed by the insertion of a network of tooling inserts that will form the faces of the roof section beams and perpendicular frames. At this point the mold is closed and injection takes place using the same resin as that incorporated into the prepregs to maintain steady hydrostatic pressure within the mold as shown in Fig. 5.43.

Once the cure is complete the final part is removed and lifted from the tooling base as shown in Fig. 5.44. Visible to the left of the tool is the heated platen press with upper and lower bolsters of welded steel that heat and clamp the tool.

There are many pros and cons when comparing this process to RTM as summarized in Table 5.4. One of the important aspects related to both processes is the part thickness control. That is possible by the use of the matched tooling, avoiding the potential thickness variation inherent in the vacuum bagging process.

Advantages of using prepreg for the SQRTM process is:

1. Availability of a wide range of materials to choose from
2. Large existing qualified database
3. Higher allowables than most RTM laminates
4. Reduce layup labor with the use of automation

When it comes to the tooling it can be very complicated. Usually the tool is made from many different parts that are joined together to create the final part shape as shown in Fig. 5.45. In this case, the tool is made to build a wing like structure with several spars across the span and a top and bottom skin. The total

FIGURE 5.43 Mold closed and resin being injected [37].

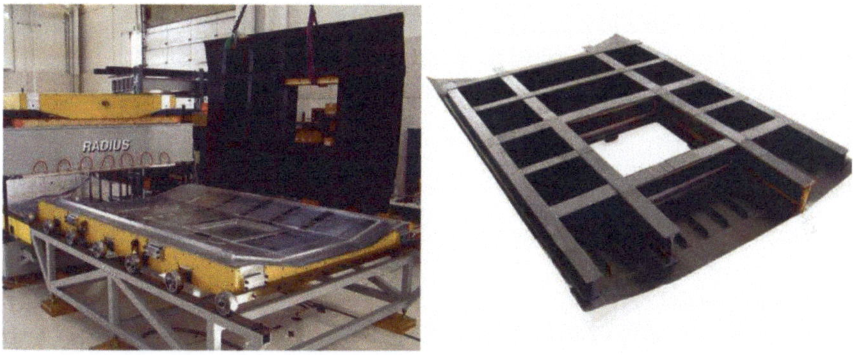

FIGURE 5.44 Part getting demolded out of the tool post cure [37].

TABLE 5.4 Comparison between RTM and SQRTM process.

RTM	SQRTM
• RTM uses dry preform • Dry fabric layups placed in tool • Vacuum is drawn on the tool • Low viscosity RTM resin is injected	• SQRTM uses prepreg layups • Layups are de-bulked under vacuum • Vacuum is drawn on the tool • Small amount of resin is injected compatible with prepreg resin

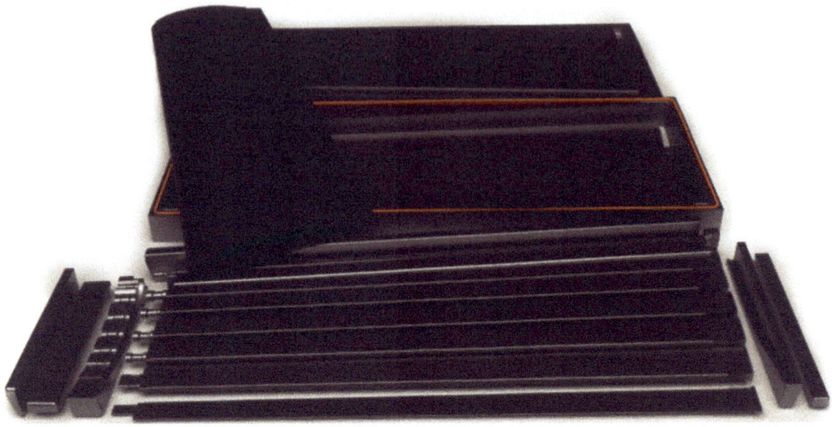

FIGURE 5.45 Set of components used to create the SQRTM/RTM tool [37].

number of part section comes to around 13 with varying levels of complexity. As you can imagine, any tolerance variation in any of these tool parts will have an adverse impact on the part performance and final shape which creates room for the ability to have more errors when it comes to tool design and fabrication. The tooling is designed to consider a number of features in order for it to perform successfully some of which are summarized below:

- Sizing for CTE
- Resin injection ports
- Vacuum ports
- Vacuum seals and part extraction and removal
- Resin flow paths
- Self-alignment features
- Critical dimension constraining

All of which are similar to all other tools. The unique thing that exist here which does not happen often is the need to evaluate how the different parts of the tool will interact with each other putting more emphasis on tolerance stack-up in this case.

5 Pultrusion

Pultrusion is a continuous process for the manufacturing of products having a constant cross section, such as beams, channels, tubing, stringers, and structural details such as tow fillers for stringers. The process is compatible with many different types of material forms including continuous strand fibers, prepreg carbon fiber or basalt fiber roving, mat, cloth, or surfacing veil. For materials that come in a dry form, it is then impregnated in a resin bath and then pulled through a steel die by a powerful tractor mechanism. The steel die consolidates the saturated reinforcement, sets the shape of the stock, and controls the fiber/resin ratio. The die is heated to rapidly debulk or cure the resin depending on the application. Many creels of roving are positioned on a rack, and a complex series of tensioning devices and roving guides direct the roving into the die. An example of the process to create the tow filler needed for hat stringer components is shown in Fig. 5.46.

The molds are typically made from hardened machined steel dies and include a preform area to do the initial shaping of the resin-saturated roving. The dies include heating which can be done using several methods such as electric or hot oil. The latest pultrusion technology uses direct injection dies, in which the resin is introduced inside the die, rather than through an external resin bath [38].

6 Compression molding

There are several types of compression molding processes that are defined by the type of material used: sheet molding compound (SMC), bulk-molding compound (BMC), and wet lay-up compression molding among others. Compression molding tooling consists of heated metal molds mounted in large hydraulic presses that are used as the layup and cure tool. Compression molding enables part design flexibility and features such as inserts, ribs, brackets and so on. The process also enables the use of automation. Good surface finishes are obtainable in this case, contributing to lower part finishing cost. Subsequent trimming and machining operations are minimized in compression molding and labor costs are low since the cure is typically much quicker compared to autoclave (30 min for compression molding vs. 8 hrs for autoclave generally speaking).

The tooling is mounted in a hydraulic or mechanical molding press and the tools are heated from 250 to 400°F depending on the material used. Based on the process, a weighed charge of molding material is placed in the open tool. The two halves of the tool are closed and pressure is applied. Depending on thickness, size, and shape of the part, curing cycles range from less than a minute to about 45 minutes. After cure, the mold is opened and the finished part is removed. This is very common process for automotive parts and is gaining more traction in aerospace application given the large interest in rate. An example of BMC process is shown in Fig. 5.47 where chopped fibers are used as the molding material. Fig. 5.48 shows a step-wise approach to compression molding using continuous prepreg material that is typically used for other autoclave parts. This is an extremely beneficial process as it cuts the cure time significantly. More work is needed to ensure that the process does not generate anomalies such as porosity or wrinkles and within the set engineering requirements. Fig. 5.49 shows a compression molding process using a rubber plug. The advantage in this case is that the rubber offers a tooling that is much cheaper than the typical tools which are usually machined or cast metal or alloy molds that can be in either single or multiple-cavity configurations which are

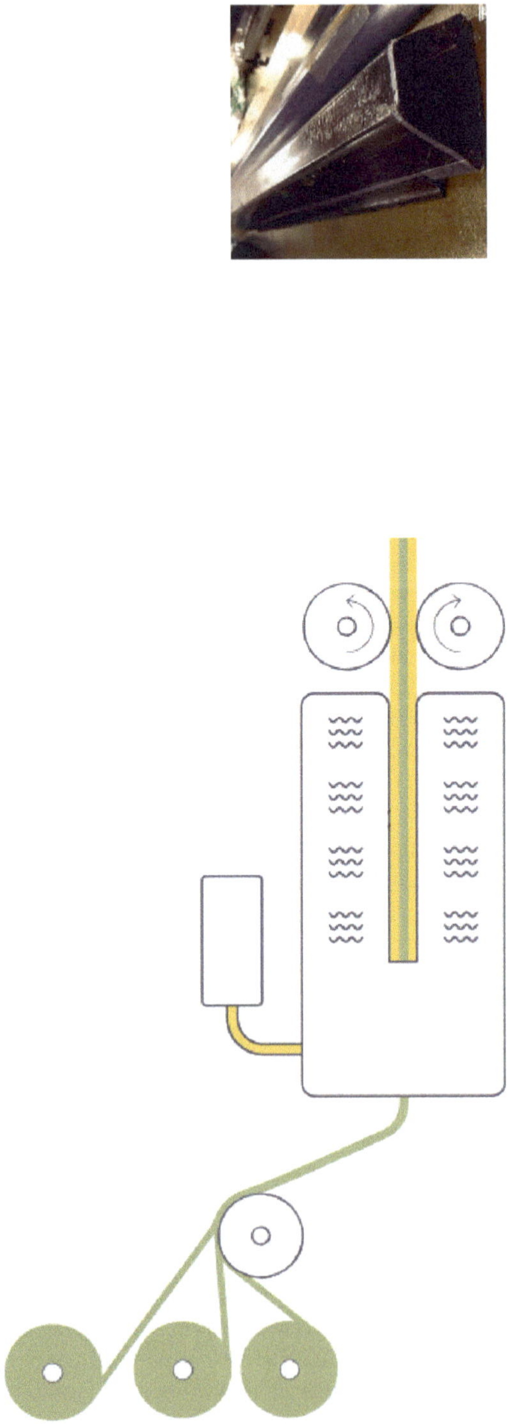

FIGURE 5.46 Example of a pultrusion process to create tow filler used for hat stringers.

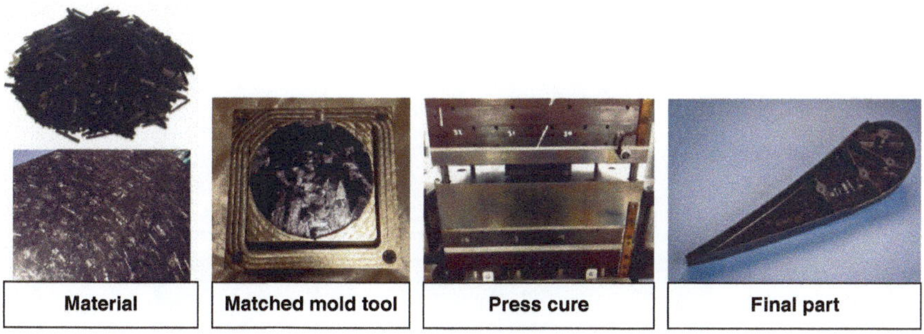

FIGURE 5.47 Bulk molding compound process steps.

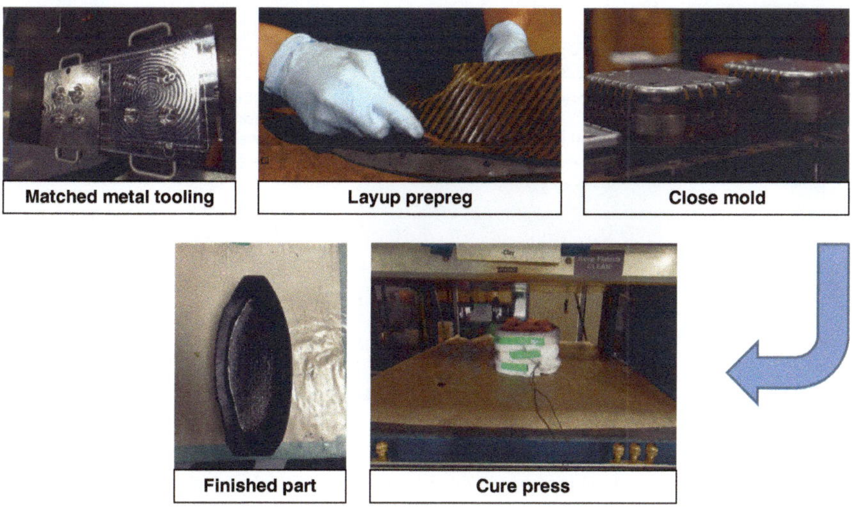

FIGURE 5.48 Example of using prepreg in compression molding.

expensive as they need to withstand the high temperature and pressure. Matched metal tools can cost 50 times as much as other type of tools and tooling in the $50,000-$500,000 range is not uncommon [39]

7 Hand layup

This is the baseline process that is currently used for majority of composite part fabrication when it comes to aerospace applications. The reason for its extensive use is the reliability it offers given its decades of implementation, ease of capturing nonconformance and performing inspection, and the ability to easily make modifications on the fly compared to all other processes. For the wing demo that was discussed in section 5.2 we used this process as the baseline. Some of the cons it offers is the slow nature of the lamination that can take up days to weeks to finalize a part and it relies heavily on the technician that is building the part.

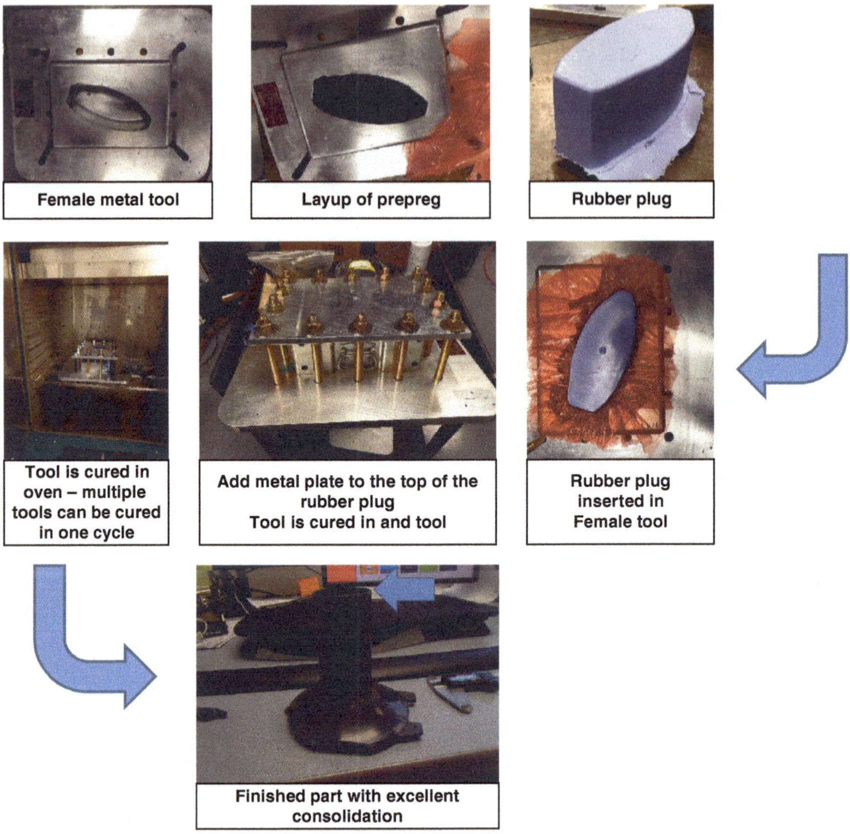

FIGURE 5.49 Example of using trapped rubber in compression molding process.

Given the slow nature of this process, the material out time needs to be tracked carefully especially for larger parts to ensure we do not exceed the time which can cause the whole part to be scrapped. The process is outlined in Fig. 5.50.

8 Composite automation

All the methods discussed so far have some sort of manual handling of the material needed. For large parts, it makes sense to consider automation to expedite the rate at which parts are made. AFP and ATL are two of the most commonly used automation processes for building composite parts that were introduced in the 1980s and have seen wide adoption especially for larger aircraft parts.

The primary difference between AFP and ATL is the width of the tow fiber used during lamination. For AFP it is between 0.25 and 1.5 inch and ATL is considered anything larger than 3 inches. Since AFP has narrower tows and can

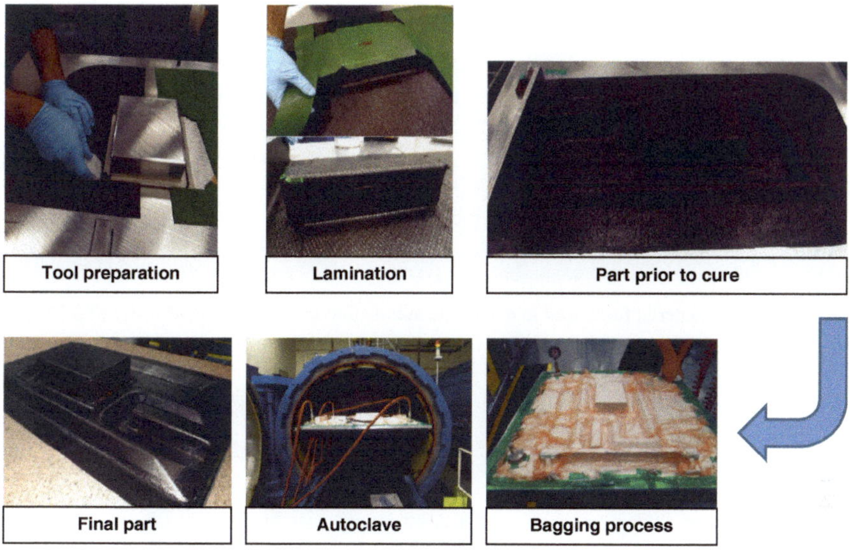

FIGURE 5.50 Hand layup process steps.

more easily be manipulated than tape, it is most effective when placing material on a curved or contoured surface. It is limited by the number of tows allowed, the total width of those tows and the length of the courses being placed. The same conventions dictate that ATL is most effective when placing large amounts of material over a relatively large flat or minimally contoured surface, and it provides high-speed laydown in such an environment. However, a large, flat surface, even if it allows for large sections of continuous tape, almost always requires the strategic placement of shorter courses of tape or fiber in a variety of different locations. Examples of the material used for AFP and ATL is shown in Fig. 5.51.

Given the tow width it has been generally assumed that ATL will be the speedier of the two processes if parts exhibit little complexity. AFP on the other hand is considered slower but is assumed to be the better choice for parts where duplication of complex contours outweighs the need for production speed. The pounds per hour of material layup is the primary quantitative value used when comparing both. For ATL is has been suggested to be around 5 lb/hr, with near-term improvement to 14 lb/hr on the horizon. For ATL it is around 30 lb/hr for a target of 60 lb/hr laydown rate [41]. Keep in mind that all these values are heavily impacted by the material and the part geometry so these should be taken as a general value. When it comes to speed, the biggest opportunities for improvement are available with the large parts - fuselage and wing skins, wing boxes, tail skins - that have been the focus of much AFP/ATL work over the past decade or so. Future of ATL/AFP could be headed more toward smaller, more complicated applications and structures, like engine cowls, nacelles, stringers,

FIGURE 5.51 Material form used in composite automation fabrication (A) AFP (B) ATL [40].

FIGURE 5.52 AFP machine [42].

and frames. The material comes in the form of creels and are attached to the machine as shown in Fig. 5.52. Depending on the machine and the number of tows needed during layup it can range from 8 - 24 creels. Fig. 5.53 shows the machine in action laying up a wing like structure.

When designing a part and tool that uses automation for layup, many different consideration need to be given that are summarized as:

FIGURE 5.53 Aerospace part lamination example using AFP [43].

- The number of tows
- The tow width
- The course centerline strategy
- Ply stagger
- The course layup direction
- Course sequencing

All these and others have an impact on the final part quality. If those variables have not been carefully considered early on during the design and manufacturing phase many anomalies could be expected as those shown in Fig. 5.54.

Those can have a very large impact on the process flow and depending on the severity, the use of hand layup might be quicker which defeats the purpose of automation at that point. Even though it is true that AFP can provide quicker part fabrication, we need to be cognizant of other factors that impact its performance as shown in Fig. 5.55. It is highly recommended that the engineering team perform a trade study for the fabrication prior to selecting to go with this option compared to any other.

When using AF/ATL during the design, stress engineers will typically determine ply angles and a minimum number of plies per angle needed for strength and stiffness. Design engineers then define the exact ply boundaries based on maximum ply drop-off rates and interface requirements. The ply boundary and fiber angle information is passed on to an AFP programmer who fills in the details such as course centerlines, tow-drop locations, course layup direction and off-part motion. Commercially available fiber placement software can be used to predict, and possibly optimize, the in-cycle time. An example of a commercially available fiber placement software is Vericut Composite Programming (VCP).

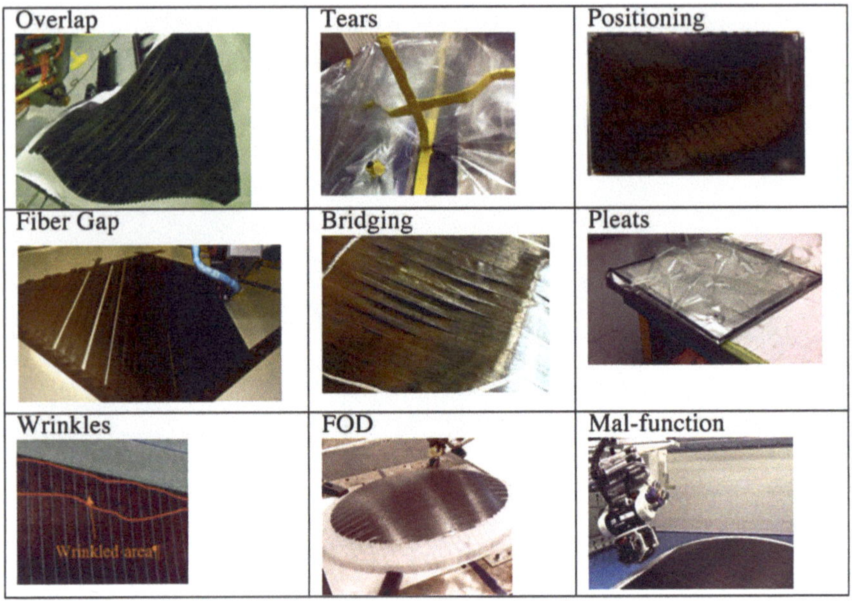

FIGURE 5.54 Typical defects in composite parts due to automation [44].

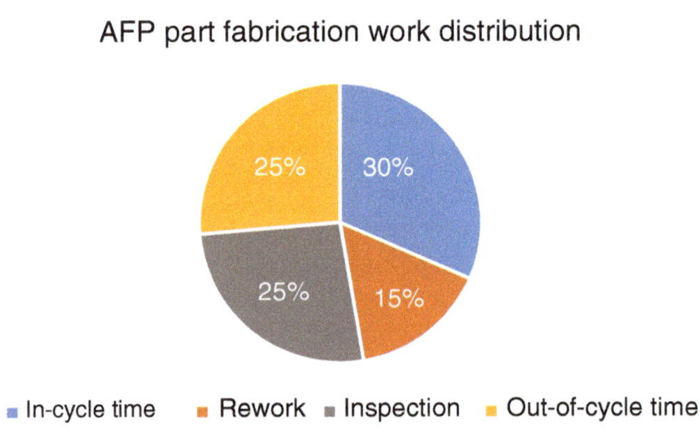

FIGURE 5.55 Factors impacting the fabrication using AFP.

Variables that need to be controlled using AFP:

- *Steering*: There needs to be a requirement set for the steering radius. The impact of the steering radius can influence the fiber direction as shown in Fig. 5.56.
- *Stroke height/roller deflection and compaction*: Need to define the requirements for roller height/deflection and compaction of the material as shown in Fig. 5.57.

FIGURE 5.56 Fiber steering [45].

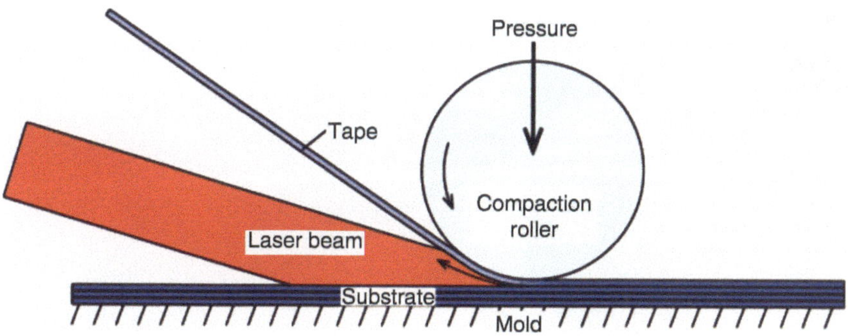

FIGURE 5.57 Schematic of compaction roller and associated variables.

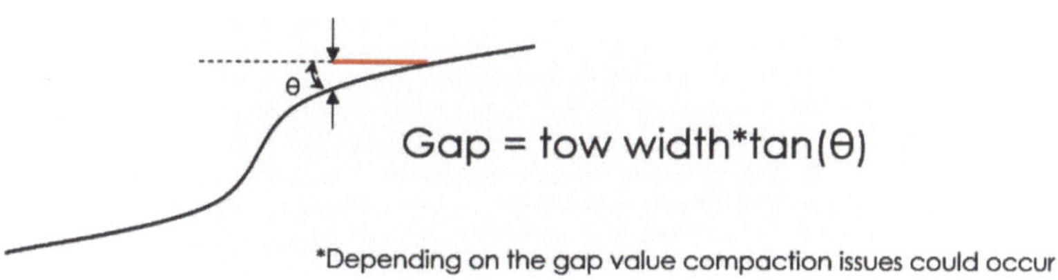

FIGURE 5.58 Normality definition when using AFP/ATL process.

- *Normality*: There needs to be a requirement set for the normality (defined by angle θ which impact the gap as shown in Fig. 5.58.

In general, wider, parallel courses with wider tows that are not steered are faster to manufacture. However, geometry might not allow courses to be parallel and follow a natural path at the same time. In addition, the fiber angles are likely to diverge from the fiber angles defined by stress, requiring courses to be both steered and nonparallel to meet fiber angle requirements.

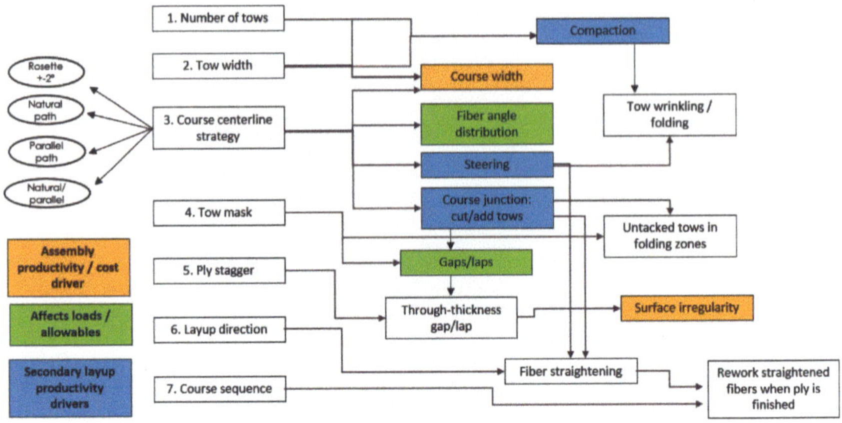

FIGURE 5.59 Flow chart of the variables impacting both AFP and ATL process.

FIGURE 5.60 Flat panel fabrication using AFP.

Fig. 5.59 shows a flow chart describing the impact of each variable on the part quality downstream. This can be used by engineers in order to determine the best path forward when selecting a manufacturing approach.

The lamination using AFP/ATL occurs one layer at a time and depending on the stacking sequence each layer is oriented to the correct direction to finally create the shape needed as shown in Fig. 5.60. This represents a simple flat

FIGURE 5.61 Wing skin example that was used to demonstrate AFP/ATL process restrictions.

panel that was built in order to understand the variability associated with a new material that we were evaluating at the time.

Another advantage of using some of the available commercial software for AFP such as VCP is the ability to inform the designer on whether the tow width in combination with the number of tow paths will be sufficient to build a part with no issues such as extreme fiber steering, gaps, laps, etc. The program also checks material conformance, visualize ply angle deviations, steering violations, roller compression, highlights excessive overlaps, and gaps. In order to demonstrate the use of the program we used a wing skin part shown in Fig. 5.61 and considered 16 parallel paths each with 0.5 inch tow width and evaluate weather this will create any manufacturing difficulties.

Fig. 5.62 shows the results of the analysis. What you are looking at is the laminate distribution at different angles (0, 45, −45, 90). Given the contour of the part in combination with the other material variables, there does not seem to be any highlights or indications from the program to expect any issues with the build and we conclude that this is a feasible set of variables to use for the build.

There are many other manufacturing techniques used in composites, but we will only focus on those discussed here. For more information the reader is referred to [46]. Next, we will go over many different stiffening techniques used in composite structures when designing parts.

9 Stiffening techniques used in composite structures

As discussed earlier, composite parts utilize many different types of stiffening features to aid in the structural integrity of the part. Here we will introduce some of the types and pros and cons of each.

9.1 Hat-stiffened part family

These are stiffeners that are used for manufacturing wings, fuselage, or any other structure that requires stability and added stiffness. These types of stiffeners

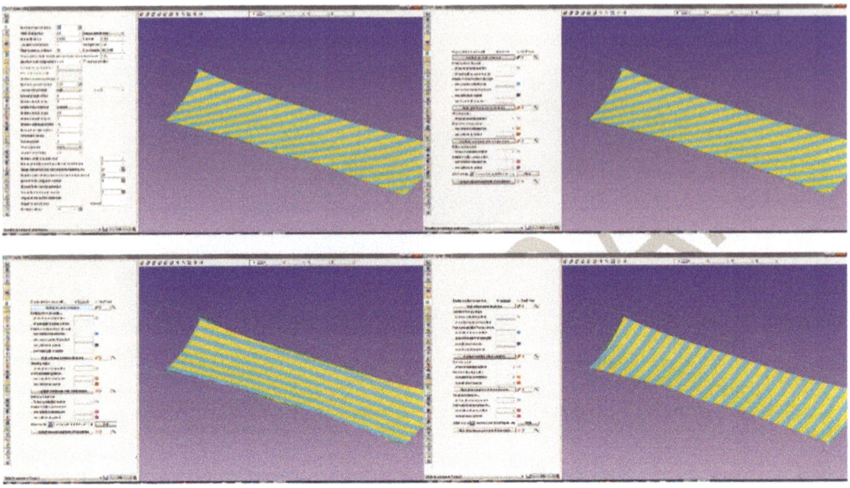

FIGURE 5.62 VCP output of the wing skin.

FIGURE 5.63 Hat stringer example.

are used for double curvature surfaces with gentle curvature (e.g., skins). An example is shown in Fig. 5.63.

The tooling approach can be done by either an OML mold with rubber mandrels or using a hollow bladder as an IML approach. Both approaches are schematically represented in Fig. 5.64.

A summary of some of the advantages and disadvantages of this stiffening approach as it relates to different disciplines is shown in Table 5.5.

9.2 Foam-stiffened part family

This type of stiffening features is used in many applications where weight is of big concern or if the part is too complex to use conventional tooling for. Note that the foam in these types of applications are "fly away" which means that it remains in the part and does not get removed. The foam also acts as a tool where

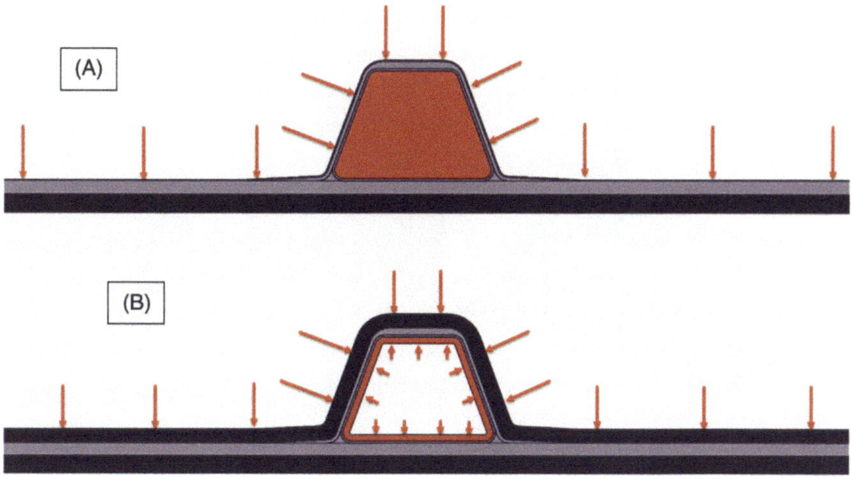

FIGURE 5.64 Tooling approaches used for hat stringers (A) OML and (B) IML.

TABLE 5.5 Advantages and disadvantages of hat stiffened structures.

Discipline	Advantage	Disadvantage
Design	• Structurally efficient design • Minimal fuel loss when used in wet bays	• Requires sealing hats between fuel tanks • Adds complexity to the design
Stress	• Point design testing is typically used to substantiate the design.	• Lack of analysis methods for this stiffener • Difficulty in analyzing the hat termination
Tooling	• OML tooling approach easy to design and use	• Higher cost • Spring back and warpage concerns due to complicated tooling
Materials and process	• Most composite prepreg systems compatible with this stiffening approach	• Wrinkling and bow wave concerns during the manufacturing process impacting allowables
Manufacturing	• Relatively easy to manufacture	• Complexity in positioning on other structures (e.g., wing skin) • Extra time needed to machine some details
Quality assurance	• The contour with this stiffening feature can be relatively easy to control	• Complex inspections required
Nondestructive inspection	• All solid laminate construction makes it easier to inspect	• Difficulty in inspecting the tow filler area

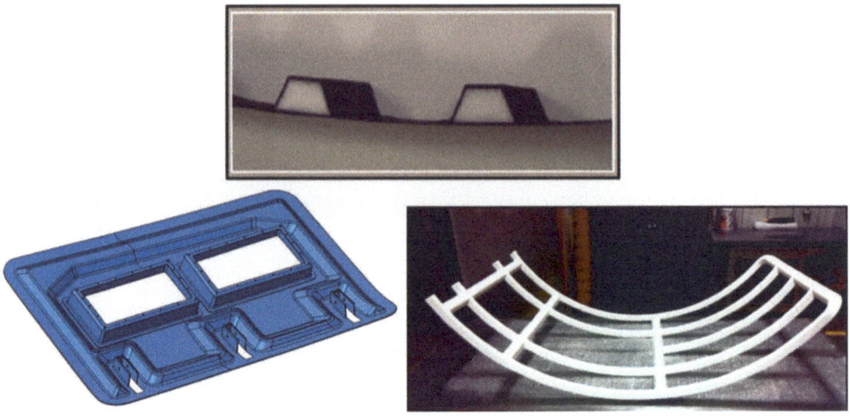

FIGURE 5.65 Foam stiffened structural parts.

the layers of a composite are laid up on top of the foam to create the derided shape as shown in some examples in Fig. 5.65.

Rohacell HERO foam [47] is an example of a typical type of foam used in such applications. Specific examples where these are used are external doors and hatch openings. Depending on the foam type, they can swell in areas of high moisture so its recommended to examine the application to ensure that no adverse impact to moisture is possible.

One of the challenges in this case can be the compaction of the plies in areas where the foam parts terminate, as we need to ensure that the layers are draped smoothly in those areas. FiberSim can be used to determine the best lamination approach and splicing including the design. An example of the results that FiberSim can provide is shown in Fig. 5.66. This is the results from a flat panel with several hat shape stiffeners. As shown, there is no large indication with the splices used that any manufacturing issues might be encountered.

A summary of some of the advantages and disadvantages for each discipline is shown in Table 5.6.

9.3 Bolted stiffeners

When using composites, it is recommended to minimize the use of bolts as much as possible given the adverse impact they have on structural integrity, weight, repair, and so on. But in certain circumstances they are inevitable which is why we still introduce them in this case as an option. The only unique aspect is that we will build the stiffener independently and then as a secondary operation fastener are used to bolt them to the remaining on the structure as shown in Fig. 5.67.

A summary of some of the advantages and disadvantages for each discipline is shown in Table 5.7.

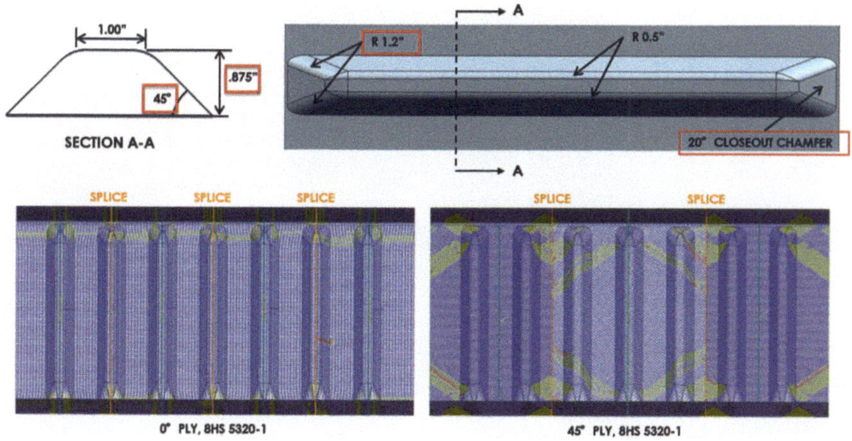

FIGURE 5.66 FiberSim results showing no distortion generated from the different splice designs considered.

TABLE 5.6 Advantages and disadvantages of foam stiffened structures.

Discipline	Advantage	Disadvantage
Design	• Relatively light weight • Flexible geometry around cutouts and curvatures	• Foam offers small nonstructural mass penalty
Stress	• Closed hat shape helps soften peel at termination	• Every configuration will require different termination thus added testing • The long-term structural impact of the foam unclear
Tooling	• Simplicity in tooling approach	• Impact of foam on spring-back and warpage unclear
Materials and process	• Most composite prepreg systems compatible with this stiffening approach	• Fluid compatibility concerns with the foam • Film adhesive may be required to be used with the foam
Manufacturing	• Relatively easy to manufacture	• Machining of the foam adds time and cost • Require definition of acceptable damage tolerances for foam
Quality assurance	• No changes in quality plans compared to other structures	• Complex inspections required specifically as it relates to foam termination areas
Nondestructive inspection	• Relatively easy to inspect around most of the structure	• New standards required depending on the final design and material selection

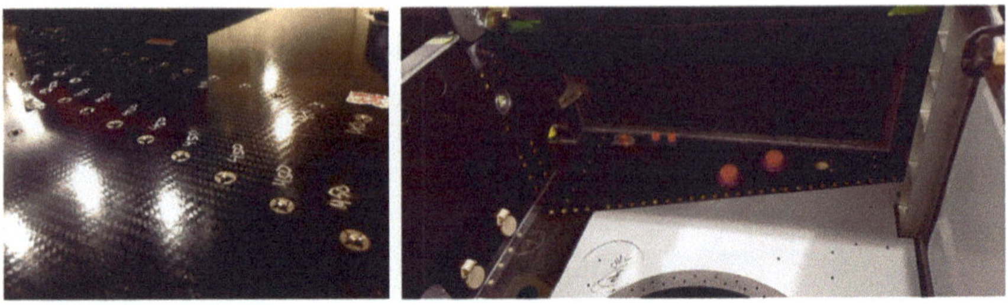

FIGURE 5.67 Bolted structure [48].

TABLE 5.7 Advantages and disadvantages of bolted stiffened structures.

Discipline	Advantage	Disadvantage
Design	• Design flexibility • Stiffeners can be tailored independent of the remaining structure	• Higher weight penalty due to added fasteners • Complexity in engineering drawing definition
Stress	• No special analysis definition required • Minimal testing needed to define special construction	• Added time needed to analyze bolted structures
Tooling	• Limited issues with thermal compensation given the secondary assembly	• May require additional tooling to support the bolting assembly process
Materials and process	• Additional material systems can be used to build different parts (e.g., compression molding for stiffeners)	• Compatibility of fasteners with the remaining of the structure
Manufacturing	• Simplified manufacturing approach • Easier to repair stiffeners individually	• Added cost and time for fastener inventory • Wet install of fasteners required
Quality assurance	• No changes in quality plans compared to other structures	• Longer inspection required due to added parts and assembly steps
Nondestructive inspection	• Relatively easy to inspect around most of the structure	• Specialty equipment may be required to inspect certain areas

9.4 Integrated parts family

This is one of the biggest advantages of using composites which is the ability to build parts integrally unitized without the need for fasteners. This minimizes the assembly time, weight, number of parts and overall cost. The main difficulty here is the tooling as it can be very hard at times to create a simple tool approach to support this type of structure. An example of an integrated unitized part is shown in Fig. 5.68. A summary of some of the advantages and disadvantages for each discipline is shown in Table 5.8.

FIGURE 5.68 Unitized composite structures [49].

TABLE 5.8 Advantages and disadvantages of unitized structures.

Discipline	Advantage	Disadvantage
Design	• Lighter weight • Reduced part count • Reduced drawing count	• Complexity in defining design details that are also manufacturing friendly
Stress	• Minimize the amount of the analysis done due to reduced part count	• Additional testing needed to inform analysis procedures for unique designs
Tooling	• Eliminates the costly assembly tooling	• Complexity in tooling design
Materials and Process	• Minimize cure time	• Lamination and producibility aspects more challenging
Manufacturing	• Minimize assembly time • Eliminates part count translating in reduced cost	• Complexity in manufacturing procedures
Quality assurance	• Reduced time needed for inspection and procedure definition	• Complicated areas to inspect and buy off
Nondestructive inspection	• Relatively easy to inspect around most of the structure	• Specialty equipment may be required to inspect certain areas

Chapter question

1. Name four different types of stiffening features used in composite design?
2. What are the factors that impact the selection of AFP/ATL process?
3. What is the main different between RTM and SQRTM?
4. What is the purpose of using sacrificial plies during composite fabrication?
5. Name one of the primary advantages of using composite structures?
6. What are the pros and cons of using compression molding?

References

[1] Sherwood, J., 2017 "U.S. Composites Manufacturing Industry, Technical Roadmap" Facilitating Industry by Engineering, Roadmapping and Science (FIBERS) Consortium.

[2] Kumara M, et al. Fractographic analysis of tensile failures of aerospace grade composites. Materials Research. 2012;15(6):990–7.

[3] Schmid T, et al. Bonding of CFRP primary aerospace structures - discussion of the certification boundary conditions and related technology fields addressing the needs for development. Composite Inter. 2015;22(8):795–808.

[4] https://aviation.stackexchange.com/questions/921/why-do-some-aircraft-have-multiple-ailerons-per-wing.

[5] http://airpigz.com/blog/2010/2/10/coolpix-airliners-airbus-a380-looking-really-big-at-oshkosh.html.

[6] http://old.seattletimes.com/html/boeingaerospace/2009565319_boeing30.html.

[7] Davies P, et al. Failure of polymer matrix composites in marine and off-shore applications. Failure Mechanisms in Polymer Matrix Composites 2012;300–36. https://www.sciencedirect.com/science/article/pii/B9781845697501500100.

[8] https://www.compositesworld.com/articles/aviation-outlook-fuel-pricing-ignites-demand-for-composites-in-commercial-transports.

[9] Rao, S., et al, 2018, "Carbon Composites are Becoming Competitive and Cost Effective" Infosys white Paper, https://www.repairerdrivennews.com/wp-content/uploads/2015/03/3-Ravi-Kumar-Infosys.pdf.

[10] Pop A, Bejinaru G. Manufacturing process and applications of composite materials. Fascicle Manag. Technol. Eng. 2010;IX(XIX). 2010, NR2.

[11] Budhe S, et al. An updated review of adhesively bonded joints in composite materials. Int. J. Adhes. Adhesiv. 2016;72:30–42. (2017).

[12] Bullen, G., 2017, "Workshop: Assembly & Joining of Composite Materials" AeroDef Manufacturing.

[13] Allen, G., Belisario, D., 2004, "Co-cured composite structures and method of making them" US Patent US6743504B1.

[14] Simpson, C. et al, 2010, "Single piece co-cure composite wing" US Patent Application US20060249626A1.

[15] Stephens, J. et al 2012, "Method and Apparatus for Co-Curing Composite Skins and Stiffeners in an Autoclave" US Patent Application US20140096903A1.

[16] Biornstad, R., 2004, "Composite barrel sections for aircraft fuselages and other structures, and methods and systems for manufacturing such barrel sections" US Patent US7527222B2.

[17] Humfeld, K., Nelson, K., 2015, "Co-curing process for the joining of composite structures" US Patent US9731453B2.

[18] Hasan, Z., et al, 2018, "Unitized Composite Structure Manufacturing System" US Patent Application.
[19] Abliz D, et al. Mixed-mode fracture toughness of co-cured and secondary bonded composite joints. Polymers and Polymer Composites 2013;21(6):341–8.
[20] Gaddikeri, K., Rao, M., 2002, "Co-curing Techniques for Integrally Stiffened Shells" Conference Paper · January 2002, https://www.researchgate.net/publication/283495184_CO-CURING_TECHNIQUE_FOR_INTEGRALLY_STIFFENED_SHELLS.
[21] https://aviationweek.com/technology/european-project-tackles-composites-assembly-hurdle-production-ramp.
[22] Huang C. Study on co-cured composite panels with blade-shaped stiffeners. Comp Part A Appl. Sci. Manufact. 2003;34(5):403–10.
[23] Kim G, et al. Manufacture and performance evaluation of the composite hat-stiffened panel. Composite Structures August 2010;92(9):2276–84.
[24] Alan Hiken, 2017, "The Evolution of the Composite Fuselage - A Manufacturing Perspective" SAE International.
[25] CMH 17, https://www.amazon.in/Composite-Materials-Handbook-CHM-17-Composites/dp/076807813X.
[26] Hasan Z. An investigation into the performance of composite hat stringers incorporating nanocomposites using a multiscale framework. J. Reinfor Plas Comp. 2014;33(15.).
[27] Hasan, Z., Chattopadhyay, A., 2013, "Thermo-Mechanical Analysis of Structural Elements Incorporating Nanocomposites" ASME 2013 International Mechanical Engineering Congress and Exposition Volume 9: Mechanics of Solids, Structures and Fluids, https://asmedigitalcollection.asme.org/IMECE/proceedings-abstract/IMECE2013/56383/V009T10A084/263154.
[28] Hasan Z, et al. Multiscale approach to analysis of composite joints incorporating nanocomposites. J. Aircraft 2014;52(1):204–15.
[29] https://www.plm.automation.siemens.com/global/en/products/nx/fibersim.html.
[30] Price, T., 1997, "Handbook: Manufacturing Advanced Composite Components for Airframes" U.S. Department of Transportation, FAA.
[31] https://www.convergent.ca/products/raven-simulation-software.
[32] http://www.radiuseng.com/net_shape_composites.
[33] https://amarketresearchgazette.com/composite-preform/.
[34] https://textechindustries.com/3d-weaving/.
[35] https://www.researchgate.net/publication/272182309_Healing_of_fatigue_delamination_cracks_in_carbon-epoxy_composite_using_mendable_polymer_stitching/figures?lo=1.
[36] https://www.highlandcomposites.com/innovation/over-braiding/.
[37] https://www.compositesworld.com/articles/sqrtm-enables-net-shape-parts.
[38] http://compositeslab.com/composites-manufacturing-processes/closed-molding/pultrusion/.
[39] http://compositeslab.com/composites-manufacturing-processes/closed-molding/compression-molding/.
[40] https://www.compositesworld.com/articles/i-want-to-say-two-words-to-you-thermoplastic-tapes.
[41] https://www.compositesworld.com/articles/atl-and-afp-signs-of-evolution-in-machine-process-control.
[42] https://www.composites.media/the-tale-of-the-tape/.
[43] https://www.electroimpact.com/Products/Composites/Overview.aspx.
[44] Denkena, B. et al., Thermographic online monitoring system for automated fiber placement processes, Composites Part B: Engineering, https://www.sciencedirect.com/science/article/abs/pii/S1359836816305285.

[45] http://pasini.ca/variable-stiffness-laminate-composite/.

[46] Campbell, F.C., Manufacturing Processes for Advanced Composites Book, https://www.sciencedirect.com/book/9781856174152/manufacturing-processes-for-advanced-composites.

[47] https://www.rohacell.com/product/rohacell/en/products-services/rohacell-hero/

[48] http://alverstokeaviation.blogspot.com/2016/12/the-alverstoke-aviation-society-guide.html.

[49] https://www.icas.org/media/pdf/Workshops/2011/ICAS%20Workshop%20presentation%2007%20Nordin.pdf.

Chapter 6

Operation of tools

Chapter outline

We have discussed the material, design, manufacturing, and quality aspects of tooling thus far and have shed light on some operational steps that will come into play whenever you deal with lamination or assembly tooling. In this chapter we will elaborate further on many other operational topics that are important to the success of any project.

Some of the operational topics discussed here can only be observed when working on the shop floor and have witnessed how they create problems. All of these are considered quality concerns and generate delays during operation. We endeavor to use many figures and schematics in this chapter to aid in understanding the topics and encourage the reader to expand their knowledge by reviewing other available information in the literature or working with the technicians on the shop floor to observe first hand some of the difficulties they face. This will only aid in performing your job as a designer, manufacturing engineer or a student to be used when entering the workforce.

1 Tool preparation

Regardless of the material used to fabricate the tool, there needs to be a release agent on the tool in order to create a barrier between the tool and part, preventing part/tool adhesion, ensure consistency of surface finish, and facilitating part removal. To be effective, a release agent must fit within the scope of the overall process and be cost effective. Application must be simple with clearly defined

Tooling for Composite Aerospace Structures. http://dx.doi.org/10.1016/B978-0-12-819957-2.00006-7

steps, drying times should be short and cure time if any must not delay the process. The product must not contain any ingredients which will affect the final product properties or post tool treatments. Finally, release must be smooth and clean with no pre-release prior to cure. There are two main types of release agents in terms of application, which are discussed further.

1.1 Internal mold release agent (IMR)

IMR is a product that is dissolved in the resin mix, which is highly soluble in the carrier solvent such as styrene. During cure the product drops out of solution and migrates to the surface due to volumetric shrinkage, pressure and temperature. When selecting an IMR agent the following criteria should be considered:

- Total solubility in the system
- Minimal effect on cure
- Minimal effect on color
- No negative influence on physical properties
- No silicone, non-stearate, and no natural waxes which would adversely affect painting or bonding adhesives
- A consistent clean release with no build-up
- Measurable reduction in cycle time

IMR's are more commonly used with mechanized processes but can be used with highly detailed hand lay-up tooling. One major advantage is consistency since the external release systems rely solely on operator skill, not always meeting the requirements of the designer or formulating chemist. Some of the disadvantages it offers is [1]

- The need for a cleaning process if bonding or painting,
- There is high defect rate (surface defects/knit line)
- Often ineffective on its own; need of external release agent
- May decrease polymer performance

Examples of such a product are Wurtz PAT 651 and Marbocote 516FC. Fig. 6.1 shows result of applying internal release agent.

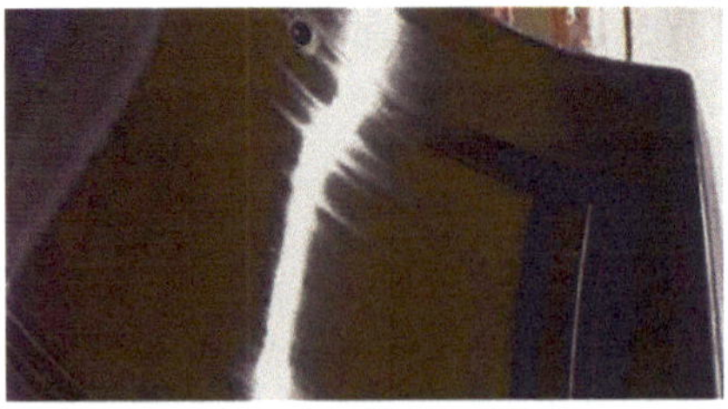

FIGURE 6.1 Illustration of the wetting of Marbocote W1151B release agent onto a carbon/epoxy tool [2].

1.2 External mold release agent (EMR)

EMRs are applied to the exterior of the tool surface and as such rely heavily on the care and attention during application to ensure successful release.

When selecting an EMR agent the following criteria are desirable:

- Easy application with no complex instructions
- No harmful solvents
- Good wetting of all surfaces, that is, should not shrink back when applied
- Surface tension should be correct for gel coat application
- Quick drying
- Easy to polish
- Any transfer to molding should be easily removed

Release agents can either be waxes, silicones, PTFE (Teflon) or soaps. In some cases, release papers are applied to the tool. These types of applications have pros and cons that are summarized below:

Advantages:

- Easy release of wide range of products
- High slip
- Easy and quick application
- Custom formulating
- Low price

Disadvantages:

- Transfer to the release surface (contamination)
- High-defect rate (surface defects / knit line)
- Cleaning process required if bonding or painting
- Release agent build-up/frequent mold cleaning

Wax-release agents are used typically in low volume manufacturing where products are individual and cycle times are not critical. Fig. 6.2 shows a tool with release agent being applied.

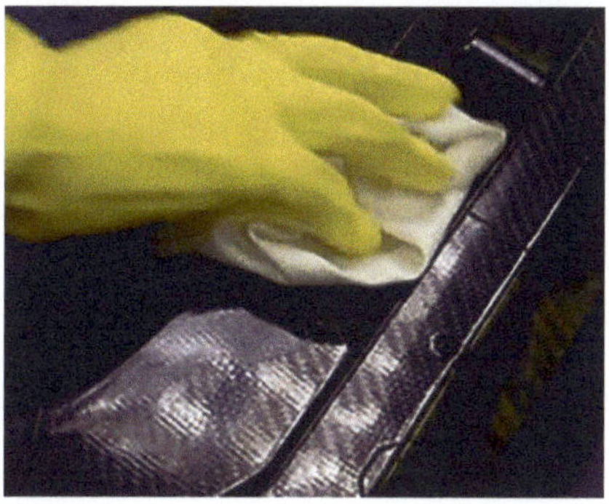

FIGURE 6.2 Release agent being wipe-applied to a carbon/epoxy composite mold [2].

Semi-permanent polymer mold-release systems enable multiple parts to be molded and released with a single application, in contrast to paste waxes, which need to be reapplied for each part.

The semi-permanent system is ideal for all contact moldings, large or small, giving exceptional surface finish with low maintenance, but without critical application procedures. In high production processes such as RTM, the use of semi-permanent systems maximizes cycle times because of their consistent release properties.

Advantages:

- Durable; multiple releases per application
- Minimal transfer; often no part cleaning required
- Reduced mold build-up and mold cleaning
- Lower defect rate
- Lower cost

Disadvantages

- Dry film; harder release/low slip
- Less user friendly; application specific
- Cure time required
- May cause pre-release if over-applied
- Stability (shelf-life)

2 Transportation and movement

One aspect that is usually overlooked when selecting a material system and design for a tool is the transportation from the fabricator to the location where it will be used. This includes the challenges associated with moving the tool around the shop floor. For smaller companies having a crane is a luxury and if there is no crane then the ability to move heavy tools is very limited. An example of a larger-heavy tool is shown in Fig. 6.3. Depending on the design, it might be altered to make the tool several parts that are connected together to make the final tool in order to ease the handling but again a trade study needs to be done to define the pros and cons and make an informed decision.

Transportation can be another detail that falls through the cracks if close attention was not directed toward understanding its limitation for the specific design considered. During my career and on several occasions we dealt with the dilemma of transportation, once for moving a large autoclave and another for an assembled aircraft. We had to work on several alternatives in those instances and decided on the best path. Fig. 6.4 shows the transportation of a large tool and/or part that will typically be the way transportation will occur using larger trucks on highways.

FIGURE 6.3 **Example of a large tool that requires a crane for movement [3].**

FIGURE 6.4 **Transportation of a large tool and/or part by ground [4].**

3 Metal bonding

In many instances, there might be a need to bond metal to composite tools or any other type of tool after one has been fabricated for the purpose of satisfying a purpose such as lifting, adding a locating feature, or handling support. If that were to occur, the engineer needs to understand the details associated with metal bonding that we briefly describe here. Metal bonding is similar in nature to secondary bonding of composites except with metal substrates instead of cured composite substrates. Metals require very stringent surface preparation including application of corrosion inhibiting primer prior to bonding to obtain long-term bond-durability at the metallic interface. Care must be taken when bonding metal to carbon as galvanic corrosion can occur in the metal substrate

To achieve an optimum bond with metals, the following guidelines are provided:

- Clean surfaces free of oils and dirt if applicable
- Refresh oxide layer with suitable process
- Chemically etch or couple to fresh oxide layer
- Apply corrosion inhibiting primer as needed
- Use appropriate adhesive for the application
- Provide uniform bondline thickness
- Provide constant clamping pressure
- Cure the adhesive to achieve structural properties

Fig. 6.5 shows a tool with a handling feature secondary bonded to the tool postfabrication.

It is important to select the important type of structural adhesive depending on the application. Most of the adhesives are thermoset polymers, commonly available in three main types, or chemistries: acrylic adhesives, epoxy adhesives, and urethane adhesives. They will not melt or change with environmental exposure, temperature or time. Acrylics and epoxies can withstand temperatures from −40 °F to +400 °F. Most urethanes are good up to 250 °F with a low-end slightly better than the others. Table 6.1 provides an overview of the application of each

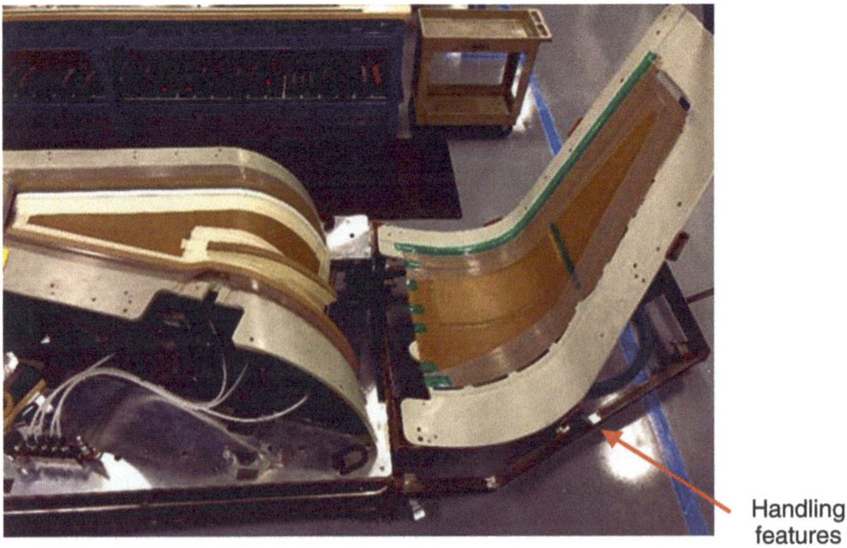

FIGURE 6.5 Tool that was modified to include a handling feature secondarily bonded [5].

TABLE 6.1 Adhesive options for metal to composite bonding [6].

	Acrylic	Epoxy	Urethane
	Pre-application phase		
Adhesive components	2	2	1 or 2
Substrate	Metals Thermoplastics Thermosets Composites	Prepared metals Rubber Thermosets Composites	Thermoplastics Rubber Thermosets Composites Primed Metals
Surface preparation			
Metals Thermosets Thermoplastics	No Yes No	Yes No No	Yes No No
Physical state	Med. liquid to paste	Med. liquid to paste pasts	Med. liquid to paste
Packaging	3 oz - 55 gal	3 oz - 55 gal	3 oz - 55 gal
	APPLICATION PHASE		
Cure temperature	Room temp. or heat	Room temp. or heat	Room temp. or heat
Working time	1 - 30 min	5 - 180 min	4 - 120 min
Handling time	2 - 60 min	2 - 12 hr	0.5 - 24 hr
Speed cure using	Mild heat/catalyst	Heat	Heat/catalyst
Flash point °F(°C)	>50 to 200 (10-93)	>200 (93)	>2O0 (93)
Humidity dependent	No	No	Yes, single-component
Mix required	No	Yes	Yes
	Post-application phase		
Shear strength	Very high	Very high	High
Peel strength	Medium	Medium	Medium
Impact strength	High	High	High
Resistance to: Moisture Chemicals UV light	 Excellent Excellent Excellent	 Very good Very good Excellent	 Very good Very good Very good
Temperature range	−40°F to 400°F (−40°C to 204°C)	−40°F to 400°F (−40°C to 204°C)	−40°F to 250°F (−40°C to 121°C)

4 Assembly challenges

There are enormous amount of challenges when it comes to the assembly of any set of parts to create a final product but when dealing with aerospace products the challenges become even larger given the tight tolerances that are needed. These tight tolerances are a big part of why the products are so reliable with minimal incidents throughout history compared to any other industries. One of those challenges when it comes to the assembly is the use of fasteners. As shown in Fig. 6.6 most of the time spent during assembly is on the drill and fill of holes. And with every hole drilled there is a high chance of creating a nonconformance (e.g., oversize hole, hole misaligned, etc.). This is why there is much interest in using unitized structures to mitigate the use of fasteners. We are not there yet as an industry so we still need to understand details related to fastening and why they are so invasive and how we can make it better in future programs and designs.

Fig. 6.7 shows examples of fastening in action, as you can see, it is a complicated process and dangerous in certain instances needing rigorous safety precautions put in place to mitigate any injury.

The process involved in fastening has many steps which are summarized in Fig. 6.8. Once the initial drilling occurs there needs to be a cleaning and deburring step where all the residual material caused by the drilling is removed and any drill coolant liquid is cleaned. Note that if the burr is not removed it can cause large fatigue concerns for metallic parts and a knockdown of around 25% to the material allowables is sometimes applied. This step is also material stack dependent, if the stack-up is only made from composites which does not generate any burrs this step can be optional. In addition, there are drilling techniques called one-up assembly [7] where the product is assembled one time—drilled, inspected, and ultimately fastened—without removal of components for

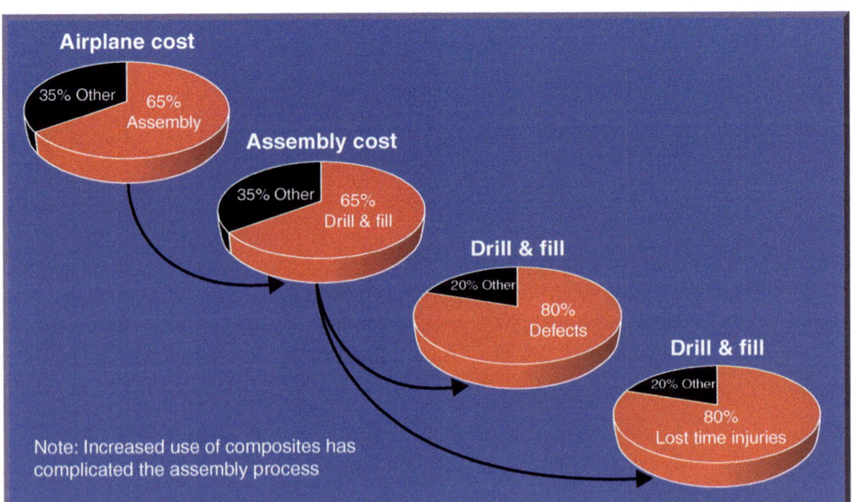

FIGURE 6.6 Chart showing the impact of hole drill and fill on assembly.

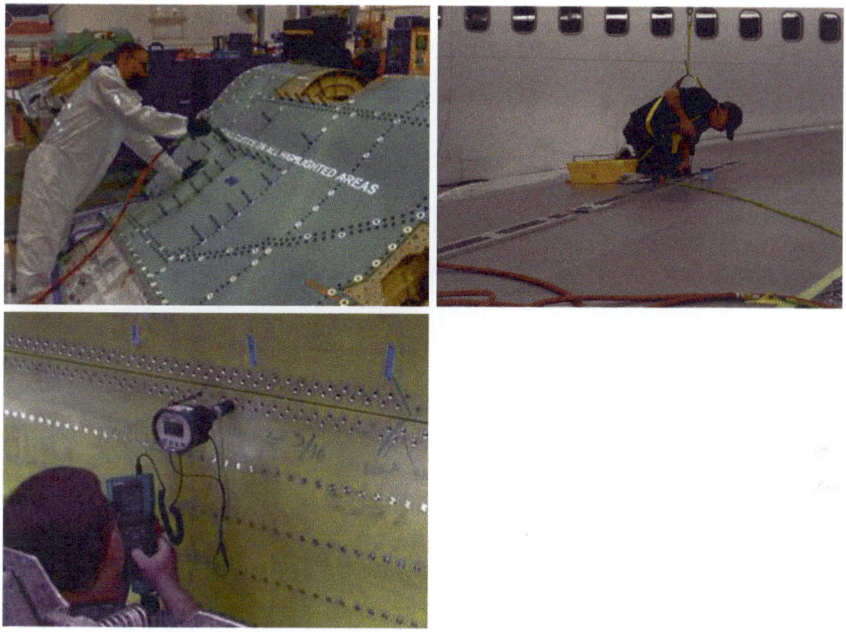

FIGURE 6.7 **Fastening process on aircraft components showing the difficulty and the time-consuming process [8].**

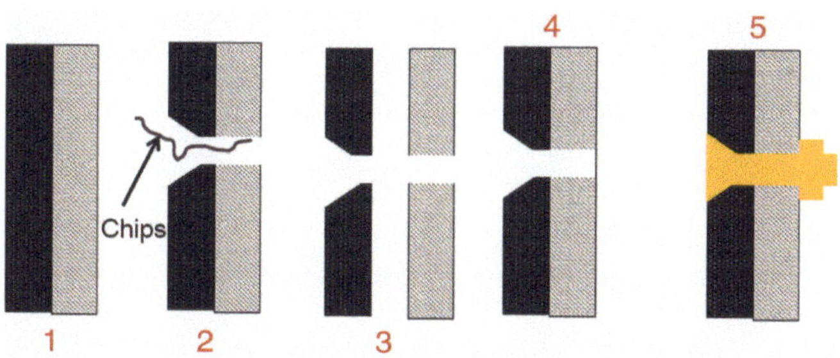

FIGURE 6.8 **Steps involved in the drilling process of aircraft components.**

deburring, cleaning, sealing, etc. This technique is preferred but will need to be qualified to ensure the process can enable this technique. This is followed by applying sealant to the fasteners and interfaces as needed. An example of a quality drilling process is shown in Fig. 6.9 that includes a combination of a composite and metallic material including a liquid ship in the middle.

In the past several decades' advances in automation for fastening have expanded rapidly and companies are adopting them more often given their reliability and speed. Examples of that is shown in Fig. 6.10 and there are many companies involved in this field such AIT, Ascent, MTores, and Amro, and many others.

During the PDR phase of any program a detailed outline of the assembly sequence needs to be thought through and discussed with the different disciplines to gain feedback and adjust the plan or changes to the design in some instances.

FIGURE 6.9 Example of a metallic to composite stack-up drilling [8].

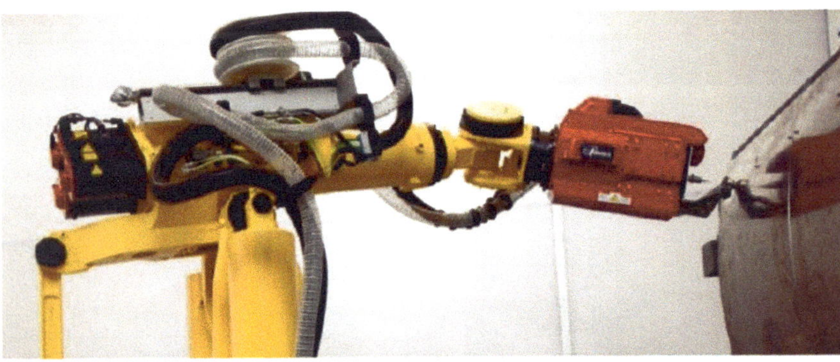

FIGURE 6.10 Automation processes used for assembling aircraft components.

The tooling approach, airframe design, manufacturing approaches and assembly plan all go hand-to-hand with each other. Starting this discussion early on in a program you are guaranteed to have the information necessary to discuss details with suppliers especially if you are using outside vendors for equipment or tooling design and fabrication. The details available will help the vendor understand the assembly philosophy and provide recommendations and accurate estimates of cost.

The plan is usually set by the tooling and design team mainly as they have the prime responsibility to design the parts and associated tooling. Depending on the company size there may be teams that are focused on the assembly process entirety and in that case, they will be the prime point of contact.

An example of an assembly flow is shown in Fig. 6.11 which is the F-35 airplane. This is considered a fairly complicated process keeping in mind that this airplane has parts built in many different companies nationally and

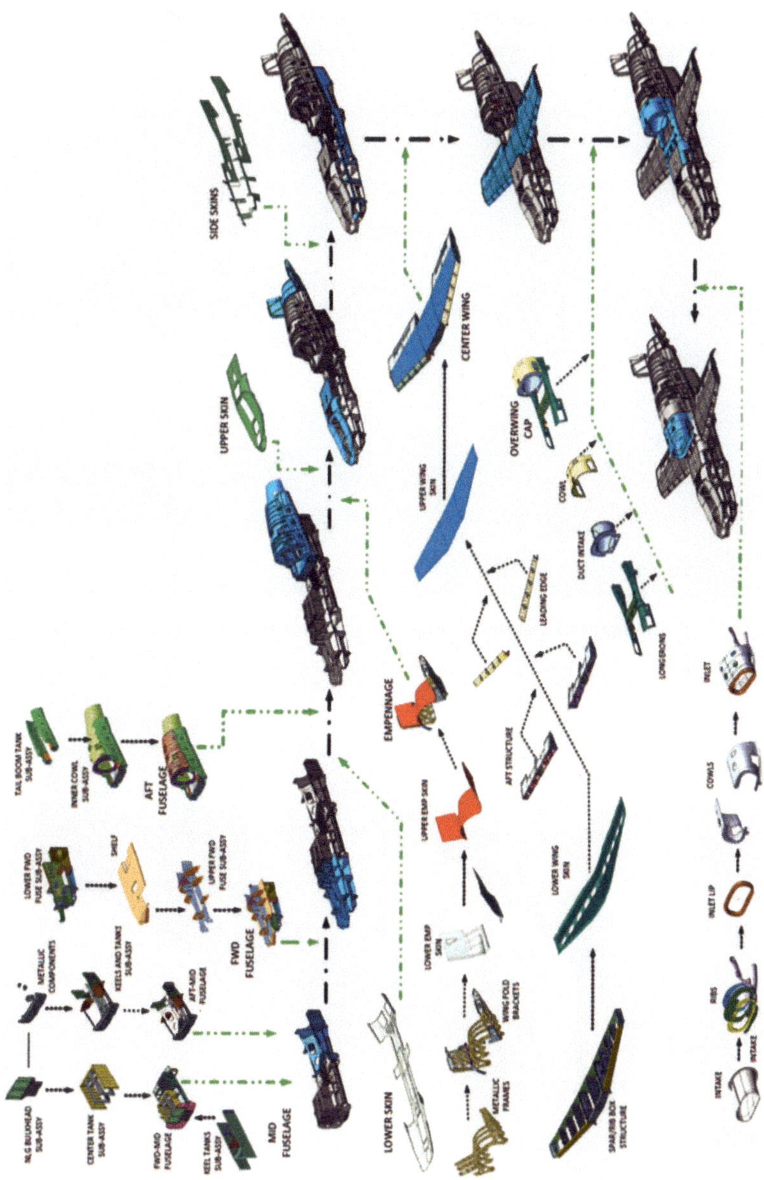

FIGURE 6.11 **Assembly flow for aircraft build [9–10].**

internationally making the logistics and supply chain complicated all impacting the assembly flows downstream.

Depending on the complexity of the product the number of assembly tools that need to be designed can range from 10–100 tools with varying level of complexity. Figs. 6.12–6.16 show different types of tools that are used for the

FIGURE 6.12 **Combination of assembly platform and ground support equipment [11].**

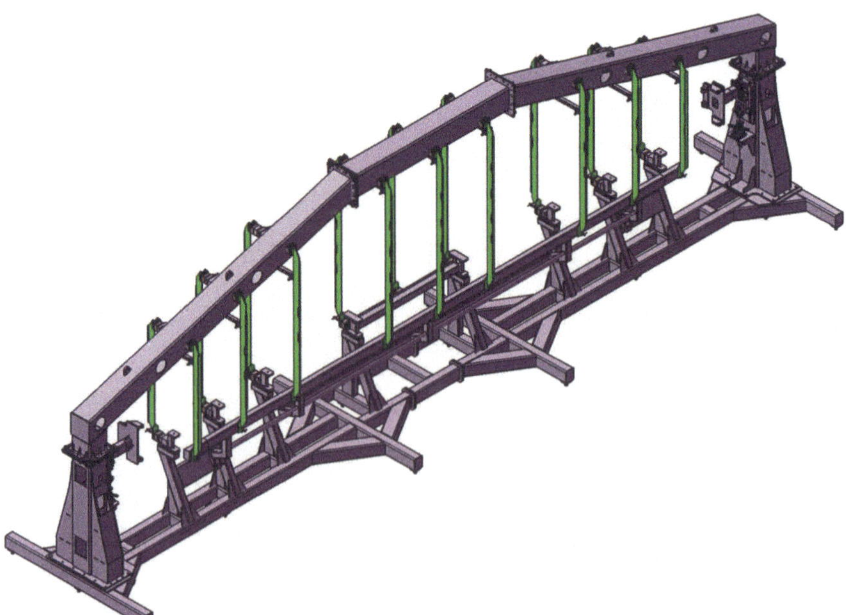

FIGURE 6.13 **Wing assembly fixture [12].**

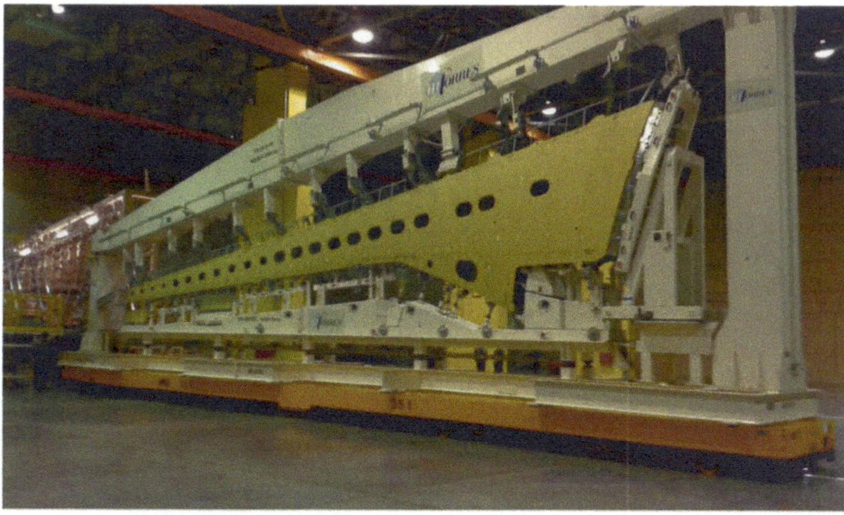

FIGURE 6.14 Assembly tools used for airplane wing [13].

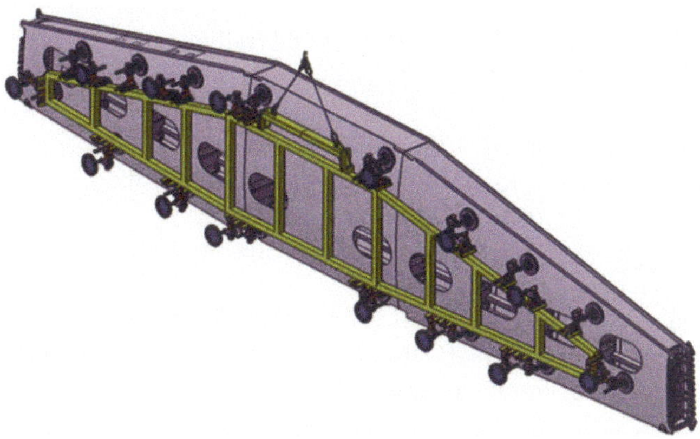

FIGURE 6.15 Wing skin end-effector for upper and lower skins.

FIGURE 6.16 Automation drilling process [14].

assembly. The tools can range for those needed to fixture it in place during the drilling process to others required to move it around in the shop and others needed for machining a part during assembly.

5 Shimming

If gaps are found during the assembly process one must either close those gaps by applying force not to exceed a certain limit or shim those areas beyond that. During the assembly process it is preferable to minimize shimming as much as possible to reduce cost of assembly and maintain quality.

In a wing for example, the shimming will occur mainly in four areas as shown in Fig. 6.17:

- Shim I: Rib to skin
- Shim IV: Rib to spar
- Shim II: Rib to stringer flange
- Shim III: Spar to skin
- Other (Major fittings near leading edge and trailing edge, and so on)

There are many reasons for why these gaps occur that include:

- Warpage of composite parts during cure
- Tolerance build up from machining of parts (skin, rib, etc.)
- Equipment measurement limitations
- IML variation of wing skin due to several factors (ply thickness, laminate compaction, cure cycle, bagging, etc.)
- Variation of equipment and tools used during assembly (e.g., overhead mechanical equipment (OHME) that is typically used to assist mechanics to transport the wing around and during assembly)

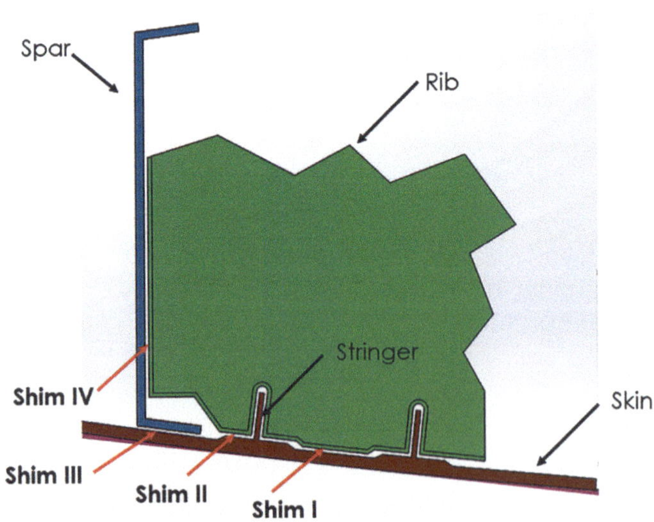

FIGURE 6.17 **Different shimming scenarios on a wing structure.**

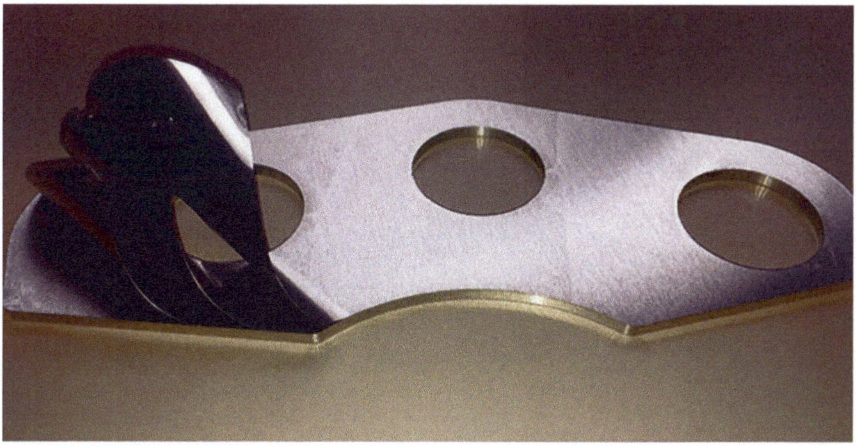

FIGURE 6.18 **Example of peel ply shims.**

Depending on the application and program, the variation to what gaps should be shimmed and which should not can be a matter of debate. For commercial airplane applications, the max gap = shim thickness = ¼ diameter of the nearest hole. For example, a ¼ inch diameter fastener, the max shimmable gap = 0.063 and anything larger shall be closed prior to using any shim.

There are many different types of shims to be use from and depending on the application and circumstance some are preferred compared to others. The use of peel shims as shown in Fig. 6.18 may reduce the difficulty of installing shims in some cases because the worker can peel off layers to accommodate shim placement. From a structural point of view, metallic solid, or peel shimming between frame members and structural subassemblies is the preferred method.

Adjacent flanges containing steps or non-flat surfaces are often filled with liquid shim, but this material has historically been limited almost entirely to gaps of 0.030" or less and forbidden at fuel boundaries, until recently [15]. Solid shims are almost twice as costly to install as common liquid shims and they can add significant weight contributions to the aircraft. The use of straight liquid shimming along with combinations of liquid (up to 0.030") and metal solid or peel shims (of varying thickness) accommodates the majority of shimming requirements on most aircraft throughout the industry. In an effort to reduce cost and weight, limitations on liquid shim applications have lessened and its use along fuel tank perimeters and possibly in gaps up to 0.060" are becoming attractive concepts. Table 6.2 shows a summary of shimming cost for different materials and methods. The use of Dynamold seems to always offer lower cost but the variation is not significant. The designer and manufacturing engineers need to consider this carefully to make the appropriate decision on the method of applying the shims and the materials used to what limit.

During assembly, fit-up forces are allowed to be applied to mating parts (e.g., rib to wing skin) through temporary fasteners as shown in Fig. 6.19. The

TABLE 6.2 Cost/weight estimation model for shimming during attachment of the JSF upper wing panel [15].

Shimming methods		Shim distribution		Material costs		Labor costs			Total cost & weight	
Liquid/dynamold to 0.030″ and Ti solid for gaps > 0.030″ in all cases		**by volume**	**by area**	**Unit cost (per lb.)**	**Total per operation**	**Estimated touch labot hours**	**Estimated touch labor costs**	**Estimated support labor costs**	**Total cost pet operation**	**Weight contribu-tion per operation**
Standard Liquid Shimming	EA-9377	20%	40%	$139 per lb.	$9,828	387	$29,553	$10,008	$49.4 K	23.5 lbs.
									EA-9377	
Liquid shimming using common practices and either EA-9377 liquid or Dynamold frozen shim material. Squeeze-out controlled by set-up bolts.	Al solid	80%	60%	$55 per lb.						
									Dynamold	
	Dynamold	20%	40%	$438 per lb.	$8,934	376	$28,735	$9,731	$47.4 K	33.7 lbs.
Laser Tracker Then Shim To Stops	EA-9377	40%	75%	″	$14,224	316	$24,136	$8,173	S46.5 K	32.3 lbs.
									EA-9377	
Laser Tracker measures variation on skin and structure faying surfaces which permits fab of shim stops. Squeeze-out controlled by stops.	Al Solid	60%	25%	″						
									Dynamold	
	Dynamold	40%	75%	″	$12,319	301	$22,958	$7,774	$43.1 K	32.7 lbs.

Tool Location Of Skin To OML	EA-B377	25%	60%	″	$12,707	330	$25,182	$8,527	$46.4 K	38.8 lbs.
									EA-9377	
Laser Tracker measures variation on skin and structure faying surfaces which permits fab of shim stops. Squeeze-out controlled by stops.	Al Solid	75%	40%	″						
									Dynamold	
	Dynamold	25%	60%	″	$11,381	318	$24,278	$8,222	$43.9 K	39.1 lbs.
Machine Skin IML To Nominal	EA-9377	100%	100%	″	$10,573	285	$21,738	87,361	$39.7 K	33.6 lbs.
									EA-9377	
Flange (faying) surfaces of skin IML are fabricated with ~0.060″ sacrificial material and then machined to nominal IML dimensions.	Al Solid	0%	0%	″						
									Dynamold	
	Dynamold	100%	100%	″	$8,667	265	$20,232	$6,851	$35.8 K	34.1 lbs.

FIGURE 6.19 Cleco's using to hold different part together with predefined forces [16].

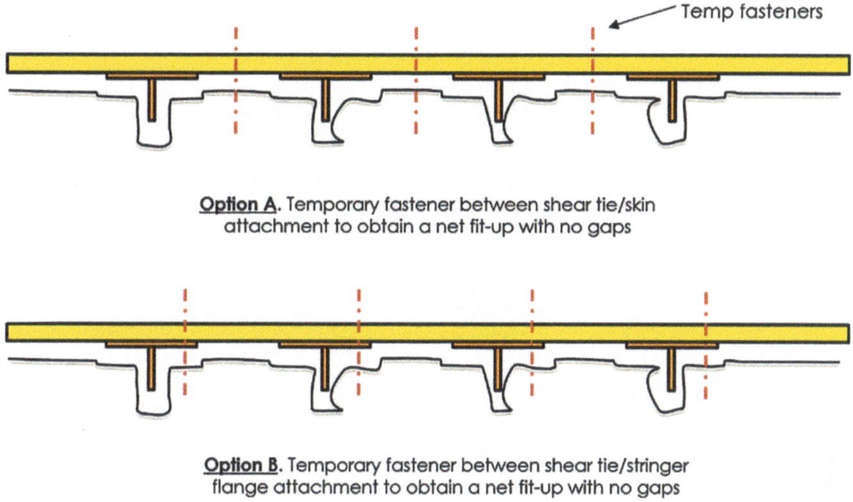

FIGURE 6.20 **Different options of how a wing can be held during the assembly process.**

load applied to the structure by those fasteners should not exceed a certain limit depending on skin thickness, rib compliance, material, and so on in order to eliminate any potential structural integrity issues (e.g., skin delamination, part cracking). The load limit to which a part can withstand can be identified via analysis. Taking a wing as an example, the temporary fasteners can be applied in many different ways as shown in Fig. 6.20. With each option comes advantages and disadvantages but one thing to remember is that the sequence of fastener application matters since you might be chasing "bubbles" if not done in a thoughtful manner and never get the part to conform and be able to close the gap.

After the allowed fit up forces are applied, mating parts that show a gap of 0.005" or less are considered "in contact" and can proceed to the permanent fastening step and they do not require further shimming other than a fay surface seal for wet tank areas. Anything above must be shimmed. A gap of 0.005" is also considered acceptable for one-up assembly drilling (0.008" might work as well but requires substantiation). A flow chart of the steps to determine the need for shimming is shown in Fig. 6.21.

Recently, there has been new approaches considered to minimize the need for manual measurement of all gaps and shimming individual areas. One approach that can be used to ensure that parts will fit up without the need for excessive shimming is building the parts with sacrificial material and machine to shape post cure. This will require the need to use computer-aided manufacturing (CAM) to shoot the parts and machine accordingly to net fit. Example of how this works is explained next:

Step 1: remove the part from the autoclave or oven (depending on the cure type) and debag the part allowing it to deform freely on the tool as shown in Fig. 6.22 which is exaggerated for illustration purposes.

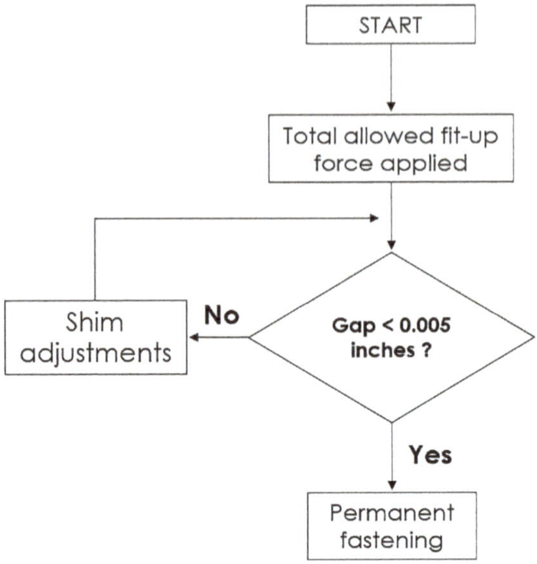

FIGURE 6.21 **Logic of using a shim during the assembly process.**

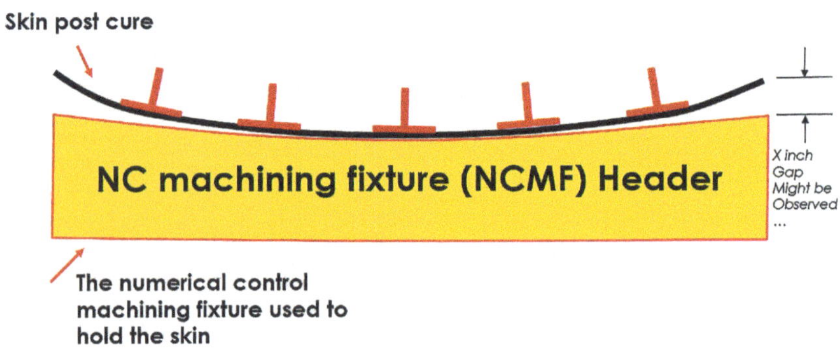

FIGURE 6.22 **Schematic representation of wing free deformation post cure.**

Step 2: apply forces to the part via temporary fasteners or though vacuum fixtures to hold the part into its nominal place ready for assembly as shown in Fig. 6.23. The forces should not exceed the allowed limit set.

Step 3: Once the part is set into its nominal location, scan the part using the appropriate metrology equipment. This will provide the exact shape of the surface including any variations during manufacturing. This step is shown in Fig. 6.24.

Step 4: The spar would have been fabricated with extra sacrificial plies on the exterior as shown in Fig. 6.25. Based on the metrology scans from the previous step those will be used in order to machine the spar surface to match.

Step 5: Similarly, the ribs will include sacrificial plies in the areas where contact is expected and using the metrology scan, we will machine the parts to ensure perfect fit as shown in Fig. 6.26.

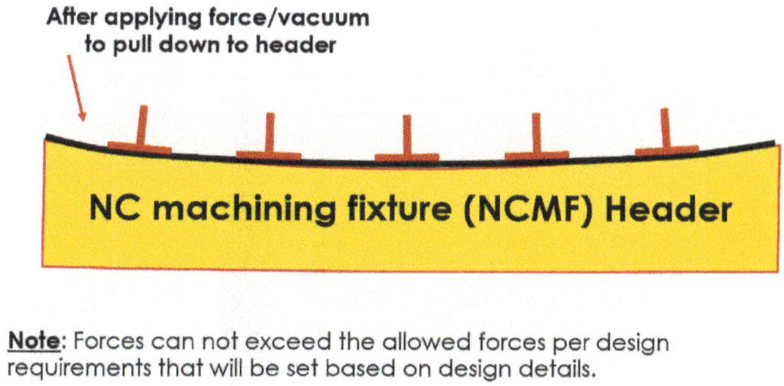

FIGURE 6.23 **Moving the part to the skin by applying external force.**

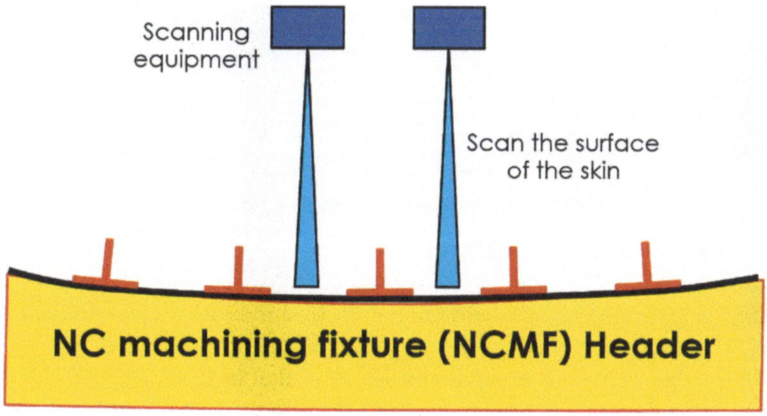

FIGURE 6.24 **The part is scanning to determine deviation from engineering definition.**

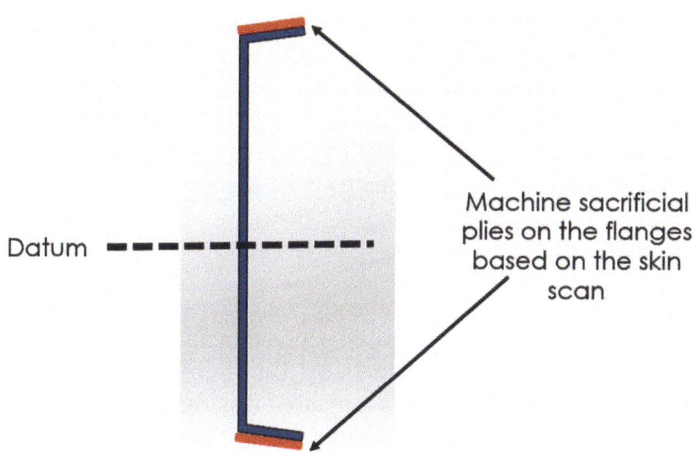

FIGURE 6.25 **Spar sacrificial plies are machined based on the data collected from the scan.**

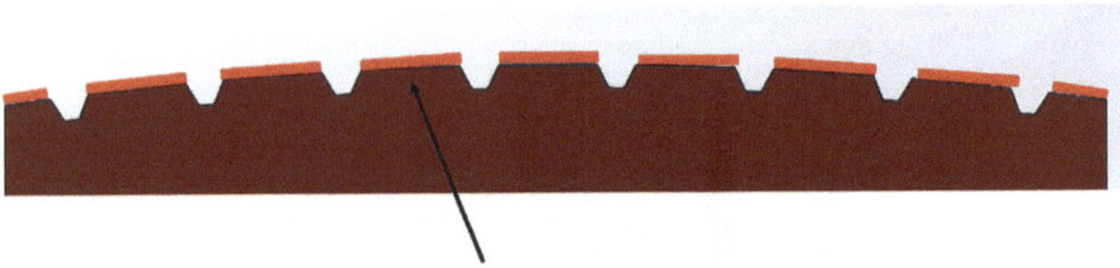

FIGURE 6.26 **Rib sacrificial plies machined per the data collected.**

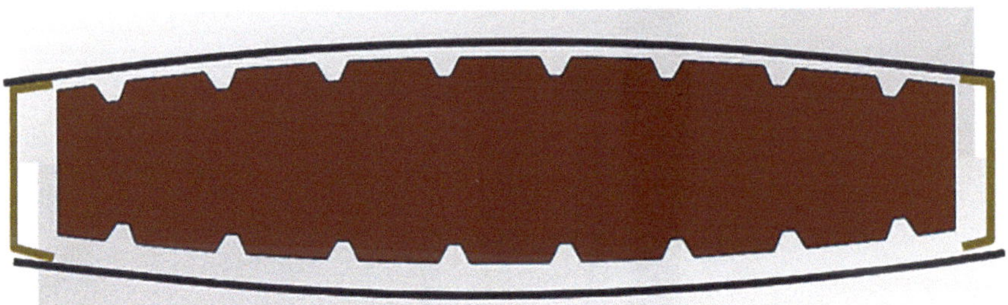

FIGURE 6.27 **Final wing components assembled.**

Step 6: Assemble all the parts together. In this case you would anticipate a perfect mating of the parts with no gaps as it was all based on actual scan of parts based on their post-production state. Note that in this case no shims were used only machining and the use of sacrificial plies. Fig. 6.27 shows the entire wing sub-assembly joined.

Some might be asking why not just use this technique all the time to eliminate the use of shims? Well the answer is that it all depends on the application, for smaller parts it may not be economical to add several plies of sacrificial material and then scan each part to assemble. It might be easier to simply use shims. For larger parts (e.g., wings) it is probably more feasible from a cost and rate perspective to use this philosophy of scanning and machining to mitigate the hundreds of shims that are anticipated to be used for each wing assembly

6 Metrology

Metrology is "the science of measurement, embracing both experimental and theoretical determinations at any level of uncertainty in any field of science and technology," as defined by the International Bureau of Weights and Measures [17]. Metrology-assisted assembly is critical to making the manufacturing process cost effective. In general, metrology-assisted assembly using advanced metrology equipment and sophisticated software results in shorter production cycles and greater cost efficiency. The past decade has seen an expansion of metrology use throughout the manufacturing process, particularly in the case of

FIGURE 6.28 **Example of coordinate measuring machine in process [18].**

aerospace. Today, metrology is widely used in production and is being adopted by design teams and other departments further upstream in the production process. The growing acceptance and application of metrology-assisted methods in aerospace manufacturing has led to the integration of metrology in everything from production, to process control and assembly, to fully automated inspection, and even troubleshooting at times.

Blue-light 3D scanners, also known as structured light, white light, or optical 3D metrology systems, use a different technology than coordinate measuring machine (CMMs) and 3D laser scanners. CMMs are programmed to collect measurements where the probe touches the part's surface as shown in Fig. 6.28. In many cases, the programming takes a long time, especially on complex geometries, thus creating a backlog. CMM data are limited to the programmed points, so locations that could possibly be suspect are overlooked, creating a long iterative process for engineers to interpret the proper corrective action.

Laser scanners on the other hand use beams of colored light to collect surface geometry. These devices that are hand-held or mounted on an articulating arm require the user to move it in paint-like strokes which can cause fatigue for the operator especially for long scan times and can cause overlapping data, especially on larger objects. Example is shown in Fig. 6.29.

Structured light 3D scanners, typically mounted on a camera stand, tripod, or robot, work like point-and-shoot cameras and require little to no programming with minimal setup. The projector quickly displays a pattern onto the part's surface, and as the pattern changes, the technology calculates X-Y-Z coordinate points

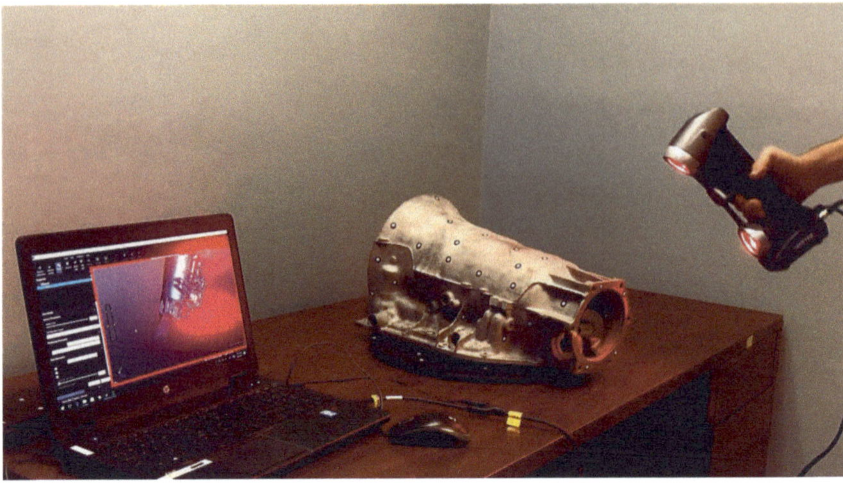

FIGURE 6.29 **Laser scanning example [19].**

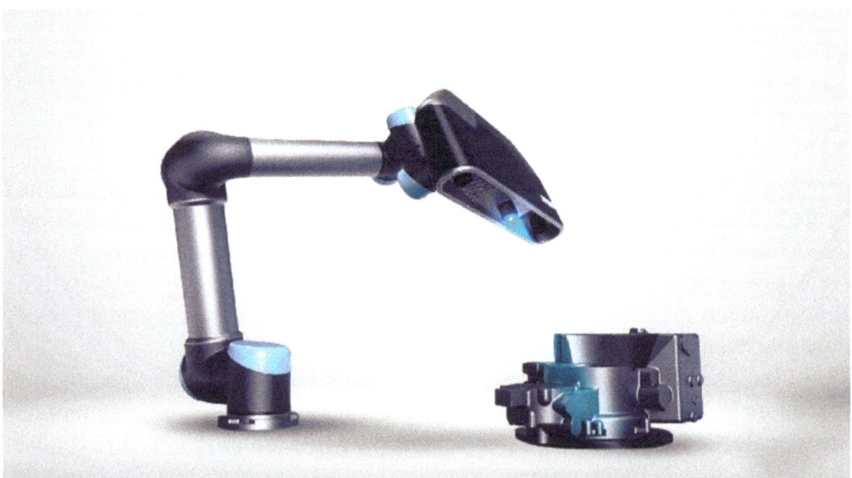

FIGURE 6.30 **Blue light technology scanning process [20].**

of an object's entire surface geometry. Algorithms are then used to automatically align each scan to create a complete digital 3D blueprint. High-quality, blue-light 3D scanners can rapidly acquire millions of accurate 3D data points per scan and interpret the results into meaningful visual feedback within the software to dramatically optimize processes. The net effect is that more surface area of the part gets measured accurately and rapidly. An example of its operation is shown in Fig. 6.30.

7 Composite part machining

We introduced the importance of composite machining of parts that is needed typically post cure in order to compensate for any anomaly the part sees during cure in order to guarantee assembly fit. It is also needed to trim part edges to

FIGURE 6.31 **Wing assembly fay surfaces that require sealant.**

create a clean surface finish for part quality. In this section, we will introduce details that the engineers need to consider when setting up machining tolerances equipment and processes.

Taking a wing as an example, and assuming that the wing will include fuel in its cavity, it is crucial to have a machined part that satisfies the set engineering requirements. The machined fay surface profile tolerances are going to be set to a profile 0.02" in order to accommodate sealant of wet structures. Faying surfaces shown in red in Fig. 6.31.

The machining is noted typically on the engineering drawing per ASME standard [21] and is shown in Fig. 6.32. The example shown informs the operator to machine the surface to a tolerance of 0.03 inches based on sides A and B.

Ensuring that you understand the tolerances is very important in any circumstance but becomes extra important when dealing with machining. In the example given we will assume a 0.02 inch total allowed profile. Profile of 0.020 inches leaves 0.010 inches room for error on either side of the designed profile. The absolute value of all errors when totaled must be $\leq$ 0.010 inch.

During a machining, we will assume that the part is within the set engineering tolerance so the different categories that the error can be applied to are caused by one or more of the following:

- Machine accuracy per the manufacturer
- Part holding which is roughly 1/3rd of the 0.010 inch tolerance
- Dynamic margin of the machine used
- Cutter/holder tolerance

The tolerance breakdown is shown in Fig. 6.33

Each category has its own limitation and applicability to the tolerance stack-up impact. Fig. 6.34 shows the different categories and the effectivity of each.

Related to the machine, there are many different machines available, but CMS is one of the largest companies in the industry that provides many different types. One of the 5 axis machines used for machining is shown in Fig. 6.35.

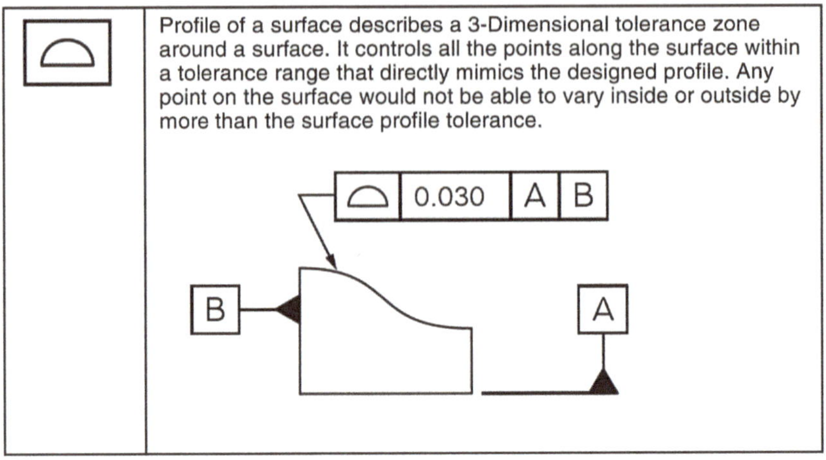

FIGURE 6.32 **Example engineering note that is required on the surface of the drawing to ensure accurate surface machining process.**

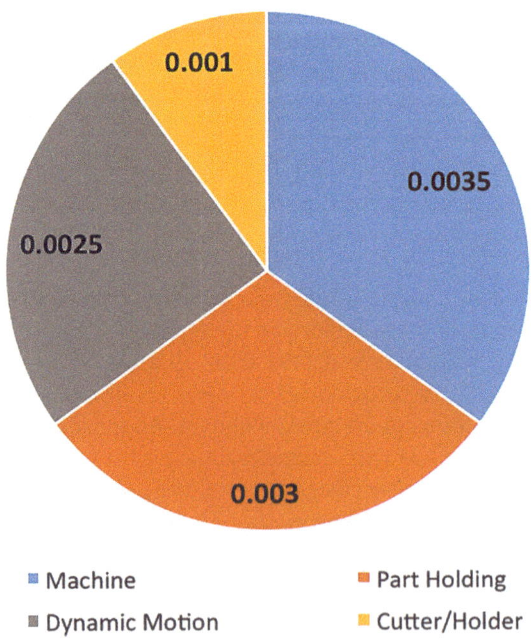

FIGURE 6.33 **Tolerance break-up among the different categories.**

A general guide for typical part machining requirements can be set as follows:

- **Surface profile 0.020 inch**: +/− 0.010 inch (focused on fay surface machining)
- **Edge trim profile 0.06 inch**: +/− 0.03 inch
- **Hole locations**: true position 0.014 inch

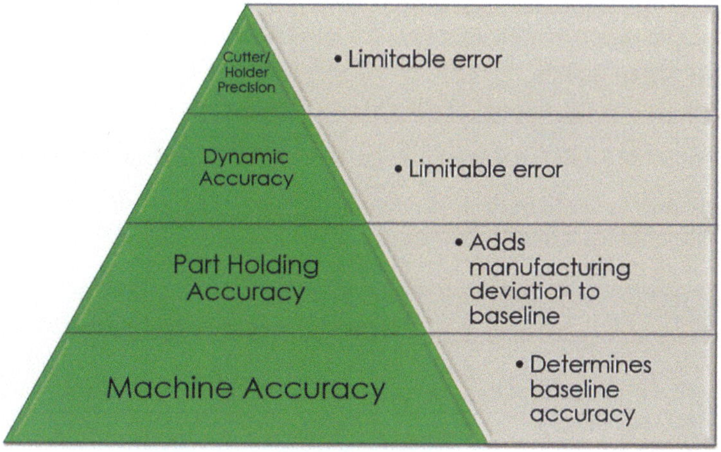

FIGURE 6.34 **The different categories impacting the accuracy of the machining process.**

FIGURE 6.35 **5-axis machine typically used for composite machining [22].**

You need to work with the equipment supplier to make sure that the machine is calibrated to eliminate that as a viable option if any issues were determined.

The dynamic accuracy has to do with the movement of the head as the machining is working which can also be calibrated and determined by the machine supplier and can be eliminated. The cutter variability depends on the type you purchase.

When it comes to equipment that is used to hold the part in place during the machining process, there are several options which include:

- Standard fixtures
- Custom fixtures
- Flexible fixtures
- Universal holding fixtures (UHFs aka POGOs)

Examples of UHF are shown in Figs. 6.36–6.38. Benefits of UHFs:

- Eliminates the need of design, manufacture, maintaining, storage, and transportation dedicated mill fixtures

FIGURE 6.36 **Standard universal holding fixture example [23].**

FIGURE 6.37 **Vertical style UHF [24].**

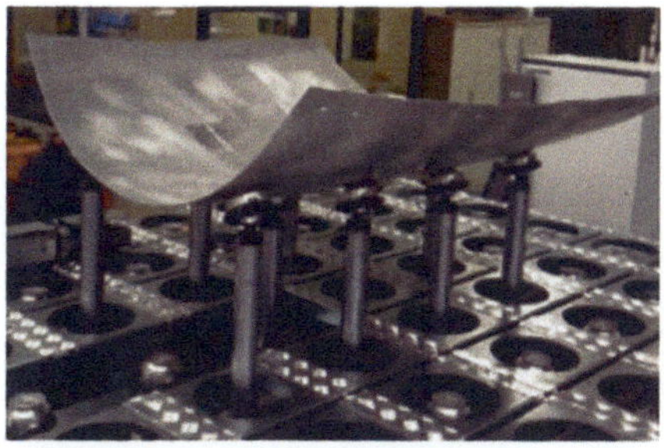

FIGURE 6.38 V-shape UHF [25].

- Quick to reconfigure from part to part
- Wide range of dimensions and shapes
- Flexible for future design changes
- Feasible for high-mix low-volume manufacturing environment
- Small footprint eliminating tool storage area

8 Nondestructive inspection (NDI) of composites

Another extremely important part of dealing with composites and tooling during the operation phase is ensuring that they are defect free to enable them to last the lifetime they are designed for. There many different types of NDI techniques that will be discussed further. For more details on this topic the reader is referred to reference [26].

8.1 Visual inspection

A visual inspection is the primary inspection method for in service inspections since most types of damage scorch, dent, cracks, abrade, or chip the composite surface, making the damage visible. Once damage is detected, the affected area needs to be inspected closer using flashlights, magnifying glasses, mirrors, and borescopes as needed. These tools are used to magnify defects that otherwise might not be seen easily and to allow visual inspection of areas that are not readily accessible. Resin starvation, resin richness, wrinkles, ply bridging, discoloration, impact damage, foreign matter, blisters, and disbonding are some of the nonconformances that can be detected with a visual inspection. These types of inspection cannot detect internal flaws in the composite, such as delamination, disbonds, and matrix cracking and other types are recommended to be used. An example of this inspection is shown in Fig. 6.39.

8.2 Audible testing

Sometimes referred to as audio, sonic, or coin tap, this technique makes use of frequencies in the audible range (10–20 Hz). This can be a very accurate method

FIGURE 6.39 **Visual inspection example that occurs typically during operation [27].**

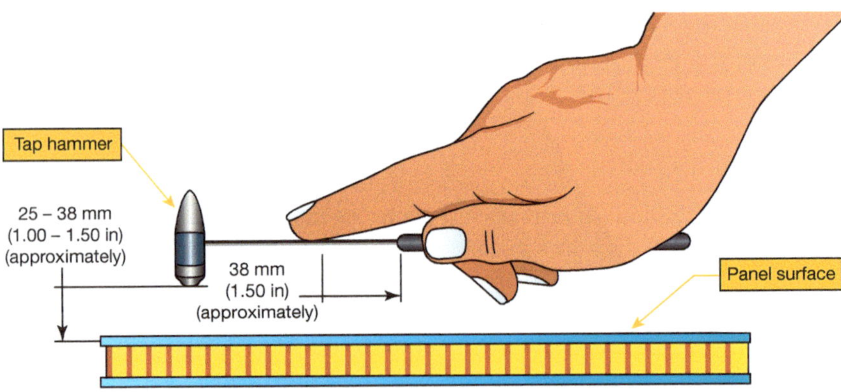

FIGURE 6.40 **Tap tester used for audible testing of composites [28].**

if used by experienced personnel, tap testing is perhaps the most common technique used for the detection of delamination and/or disband especially for core stiffened composites. The method is accomplished by tapping the inspection area with a solid round disk or lightweight hammer-like device and listening to the response of the structure to the hammer as shown in Fig. 6.40. A clear, sharp, ringing sound is indicative of a well-bonded solid structure, while a dull or thud-like sound indicates a discrepant area.

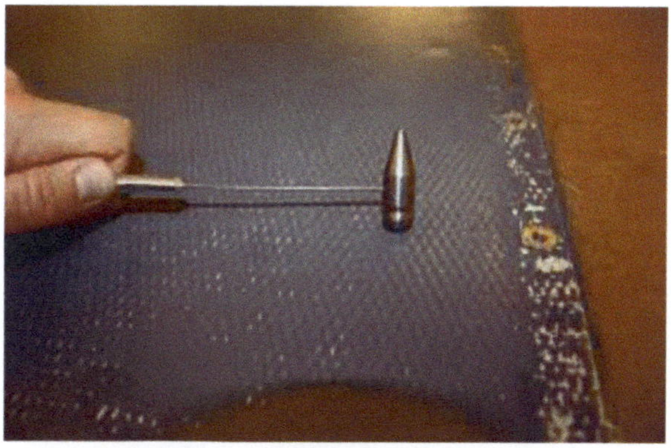

FIGURE 6.41 **Tap test occurring on a composite part [29].**

The tapping rate needs to be rapid enough to produce enough sound for any difference in sound tone to be apparent to the ear. Tap testing is effective on thin skin to stiffener bondlines, honeycomb sandwich with thin face sheets, or even near the surface of thick laminates, such as rotorcraft blade supports. This inspection should be accomplished in as quiet an area as possible and by experienced personnel familiar with the part's internal configuration. This method is not reliable for structures with more than four plies. It is often used to map out the damage on thin honeycomb face sheets. Example of this is shown in Fig. 6.41.

8.3 Ultrasonic inspection

Ultrasonic inspection is the most commonly used method to detect internal damage in composites structures and has proven to be a very useful tool for the detection of internal delaminations, voids, or inconsistencies in composite components not otherwise discernable using visual or tap methodology. There are many ultrasonic techniques; however, each technique uses sound-wave energy with a frequency above the audible range

A high-frequency (usually several MHz) sound wave is introduced into the part and may be directed to travel normal to the part surface, or along the surface of the part, or at some predefined angle to the part surface as shown in Fig. 6.42. You may need to try different directions to locate the flaw. The introduced sound is then monitored as it travels its assigned route through the part for any significant change. Ultrasonic sound waves have properties similar to light waves. When an ultrasonic wave strikes an interrupting object, the wave or energy is either absorbed or reflected back to the surface. The disrupted or diminished sonic energy is then picked up by a receiving transducer and converted into a display on an oscilloscope or a chart recorder. The display allows the operator to evaluate the discrepant indications comparatively with those areas known to

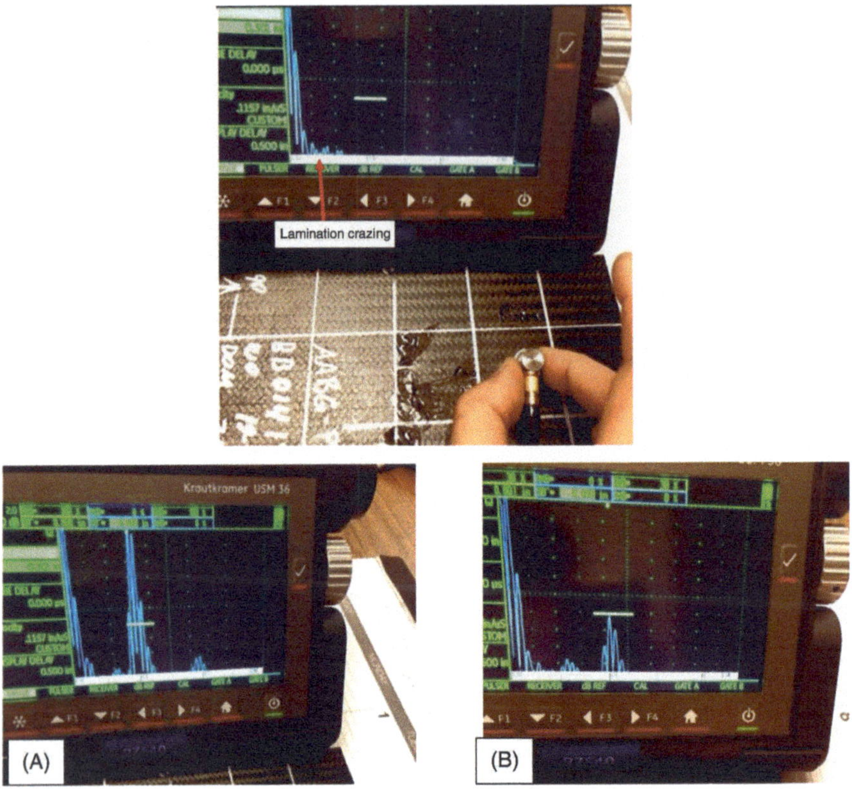

FIGURE 6.42 Ultrasonic process definition showing the variability in the signal comparing (A) part with no indication that has 100% probe surface feedback (B) another with indication and a 40% probe surface feedback indicating [28].

be good. To facilitate the comparison, reference standards are established and utilized to calibrate the ultrasonic equipment.

The repair technician must realize that the concepts outlined here work fine in the repetitive manufacturing environment but are likely to be more difficult to implement in a repair environment given the vast number of different composite components installed on the aircraft and the relative complexity of their construction. The reference standards would also have to take into account the transmutations that take place when a composite component is exposed to an in-service environment over a prolonged period or has been the subject of repair activity or similar restorative action.

The four most common ultrasonic techniques are discussed further.

8.3.1 Through transmission ultrasonic inspection

Through transmission ultrasonic inspection uses two transducers, one on each side of the area to be inspected. The ultrasonic signal is transmitted from one

FIGURE 6.43 **Pulse echo ultrasonic inspection equipment typically used for composite structures [30].**

transducer to the other transducer. The loss of signal strength is then measured by the instrument. The instrument shows the loss as a percent of the original signal strength or the loss in decibels. The signal loss is compared to a reference standard. Areas with a greater loss than the reference standard indicate a defective area.

8.3.2 Pulse echo ultrasonic inspection

Single-side ultrasonic inspection may be accomplished using pulse echo techniques. In this method, a single search unit is working as a transmitting and a receiving transducer that is excited by high voltage pulses. Each electrical pulse activates the transducer element. This element converts the electrical energy into mechanical energy in the form of an ultrasonic sound wave. The sonic energy travels through a Teflon® or methacrylate contact tip into the test part. A waveform is generated in the test part and is picked up by the transducer element. Any change in amplitude of the received signal, or time required for the echo to return to the transducer, indicates the presence of a defect. Pulse echo inspections are used to find delaminations, cracks, porosity, water, and disbonds of bonded components. Pulse echo does not find disbonds or defects between laminated skins and honeycomb core. Fig. 6.43 is an example of an equipment used for this inspection.

8.3.3 Ultrasonic bond tester inspection

Low-frequency and high-frequency bond testers are used for ultrasonic inspections of composite structures. These bond testers use an inspection probe that has one or two transducers. The high-frequency bond tester is used to detect delaminations and voids. It cannot detect a skin to honeycomb core disbond or porosity. It can detect defects as small as 0.5-inch in diameter. The low-frequency bond tester uses two transducers and is used to detect delamination, voids, and skin to honeycomb core disbands. This inspection method does not detect which side of the part is damaged and cannot detect defects smaller than 1.0-inch. Fig. 6.44 is an example of an equipment used for this inspection.

FIGURE 6.44 **Example of a bond tester on composite wing structure [31].**

8.4 Radiography

Radiography, often referred to as X-ray, is a very useful NDI method because it essentially allows a view into the interior of the part. This inspection method is accomplished by passing X-rays through the part or assembly being tested while recording the absorption of the rays onto a film sensitive to X-rays. The exposed film, when developed, allows the inspector to analyze variations in the opacity of the exposure recorded onto the film, in effect creating a visualization of the relationship of the component's internal details. Since the method records changes in total density through its thickness, it is not a preferred method for detecting defects such as delaminations that are in a plane that is normal to the ray direction. It is a most effective method, however, for detecting flaws parallel to the X-ray beam's centerline. Internal anomalies, such as delaminations in the corners, crushed core, blown core, water in core cells, voids in foam adhesive joints, and relative position of internal details, can readily be seen via radiography. Most composites are nearly transparent to X-rays, so low-energy rays must be used. Because of safety concerns, it is impractical to use around aircraft. Operators should always be protected by enough lead shields, as the possibility of exposure exists either from the X-ray tube or from scattered radiation. Maintaining a minimum safe distance from the X-ray source is always essential.

8.5 Thermography

Thermal inspection comprises all methods in which heat sensing devices are used to measure temperature variations for parts under inspection. The basic principle of thermal inspection consists of measuring or mapping of surface temperatures when heat flows from, to, or through a test object. All thermographic techniques rely on differentials in thermal conductivity between normal, defect-free areas, and those having a defect. Normally, a heat source is used to elevate the temperature of the part being examined while observing the surface heating effects. Because defect free areas conduct heat more efficiently than areas with defects, the amount of heat that is either absorbed or reflected indicates the quality of the bond. The type of defects that affect the thermal properties include debonds, cracks, impact damage, panel thinning, and water ingress into composite materials and honeycomb core. Thermal methods are most effective for thin laminates or for defects near the surface.

8.6 Neutron radiography

Neutron radiography is a nondestructive imaging technique that is capable of visualizing the internal characteristics of a sample. The transmission of neutrons through a medium is dependent upon the neutron cross sections for the nuclei in the medium. Differential attenuation of neutrons through a medium may be measured, mapped, and then visualized.

The resulting image may then be utilized to analyze the internal characteristics of the sample. Neutron radiography is a complementary technique to X-ray radiography. Both techniques visualize the attenuation through a medium. The major advantage of neutron radiography is its ability to reveal light elements such as hydrogen found in corrosion products and water.

Chapter questions

1. Name two types of release used for tool surface preparation?
2. Why does a structure need to be disassembled and cleaned after drilling process especially when metallics exist in the stack-up?
3. Name three types of adhesive bonding used during secondary bonding operations?
4. Name three different NDI techniques?
5. What are the variables that impact the machining accuracy for composite structures?
6. What is the acceptable gap limitation used to shim a part?

References

[1] http://utsrus.com/documents/seminary_doklady/marbo.pdf.
[2] https://www.materialstoday.com/composite-applications/features/new-developments-in-release-agent-technologies/.
[3] https://pcminnovation.com/molds-aerospace-composites/.

[4] https://www.customcomposite.com/portfolio-item/facility-and-crew-projects/.
[5] Shape Distortion Analysis of a Complx Shaped Wing Skin Section, June 2015. Available from: https://www.researchgate.net/publication/278933651_Shape_Distortion_Analysis_of_a_Complx_Shaped_Wing_Skin_Section.
[6] http://files.lord.com/pdf/44/SG1018_LORDStructuralAdhesives.pdf#page=4.
[7] https://www.electroimpact.com/WhitePapers/2008-01-2297.pdf.
[8] Workshop: Assembly & Joining of Composite Materials. Available from: https://www.aerodefevent.com/wp-content/uploads/2017/03/AeroDef-2017-Assembly-and-Joining-of-Composite-Materials-Workshop.pdf.
[9] http://www.infohow.org/wp-content/uploads/2012/11/F-35-Manufacturing-Sequence.jpg.
[10] https://www.assemblymag.com/articles/90231-f-35-scrap-rate-at-16-percent.
[11] https://www.ainonline.com/aviation-news/air-transport/2016-07-05/boeing-confronts-challenges-777-automation-push.
[12] http://www.aint.com/tooling/capabilities.
[13] http://www.ntew.com/solutions/.
[14] https://finance.yahoo.com/news/happens-boeing-robots-team-humans-091547898.html.
[15] Shimming During JSF Wing Carry Through Skin Attachment. Available from: Randy, L. (1999), "Evaluation of shimming options with applications to JSF", Lockheed Martin Corporation.
[16] https://www.aerotoons.com/blog/elevators.htm.
[17] Nanostructured Materials: Metrology A. Jorio, M.S. Dresselhaus, In: Encyclopedia of Materials: Science and Technology, 2010.
[18] https://www.renishaw.com/en/renishaw-to-highlight-engineering-solutions-for-the-aerospace-and-automotive-industries-at-advanced-engineering-2013--21669.
[19] https://metrology.news/3d-laser-scanning-technology-keep-aging-aircraft-flying/.
[20] https://www.indiamart.com/proddetail/faro-cobalt-blue-light-scanner-19597660691.html.
[21] ASME Y14.5-2009. Available from: https://www.asme.org/codes-standards/find-codes-standards/y14-5-dimensioning-tolerancing.
[22] http://www.mimtec.co.uk/large-5-axis-machine-centre/.
[23] https://www.mtorres.es/en/aeronautics/products/metallic/torrestool.
[24] http://www.aeromachinery.com/DufInd.htm.
[25] https://www.aeroexpo.online/prod/mtorres-disenos-industriales-sau/product-169232-23340.html.
[26] Mix P.E., Introduction to Nondestructive Testing: A Training Guide, 2nd Edition. Available from: https://www.wiley.com/en-us/Introduction+to+Nondestructive+Testing%3A+A+Training+Guide%2C+2nd+Edition-p-9780471420293.
[27] https://www.miat.edu/2017/10/aircraft-pre-flight-checklist-aircraft-safety-procedure/.
[28] http://www.flight-mechanic.com/nondestructive-inspection-ndi-of-composites/.
[29] https://www.niar.wichita.edu/niarworkshops/Portals/0/GalellaWrkshp.pdf.
[30] https://www.aimil.com/products/pile-echo-tester-pet.
[31] https://blog.boomsupersonic.com/ultrasound-inspection-the-first-step-in-wing-bend-tests-780256916341

Chapter 7

Modeling and simulation

Chapter outline

This chapter will discuss one of the important topics that have gained additional traction in recent years for use in composite materials tooling and fabrication. The topic of simulation and modeling itself has been ongoing for more than half a century but the primary use was for structural analysis and limited work was devoted to the topic of cure simulation and the impact on tooling analysis. As part of this chapter we hope to educate the reader on different software available that can be utilized for the purposes of virtual analysis and provide some examples of how to use this simulation for real life aerospace applications. *Appendix* will also provide a step by step example of how to conduct such a simulation using some of the tools we discuss here and is meant to serve as a guideline for future work.

Let us start by discussing some of the software's that are available and can be leveraged for such analysis. Fig. 7.1 shows some of those available tools, although this is not a comprehensive list, it sheds light on many of the available systems that are arguably mostly used in the aerospace field.

Creo, NX, Catia, and many others are used as the primary CAD software's that offer the first phase of any simulation which is the need for a model of the product that requires analysis. There are plenty of books that discuss these types of software's, but we want to focus here on the process modeling of composites, which is the unique aspect here.

1 Process modeling prediction software

We start by introducing four specific software's that engineers in the aerospace field use in order to model composite materials processing. Table 7.1 has a summary of those software's and some of the aspects each one exhibits.

Tooling for Composite Aerospace Structures. http://dx.doi.org/10.1016/B978-0-12-819957-2.00007-9

FIGURE 7.1 **Simulation software used for analysis of composite structures and associated tooling.**

1.1 FiberSim

FiberSim is used during the engineering process in order to generate a dataset of plies that are optimized for a specific part based on geometry to minimize defects such as wrinkles, overlaps, and gaps during the lamination process. Once a data set is generated the plies are cut to the shapes needed and laminated on the tool with the aid of laser projection. Fig. 7.2 shows an outline of the process. The software is also able to predict any defects based on geometry limitations but those thresholds are set based on inputs to the software that can be based on experience or a set of unique testing and not based on any physical model. It is highly recommended that this process be used during fabrication as a measure to ensure part quality overall.

1.2 Composite fiber modeler

Similar to FiberSim, composite fiber modeler is used to generate file datasets for lamination but interfaces with Catia CAD software rather than NX. Using this software we are also able to determine any anomalies that might exist during fabrication based on the geometric limitations that must be set using the user, which are based on experience or testing. Note that based on the material used and fiber form (e.g., unidirectional, plain weave, etc.) those limitations will change and need to be adjusted accordingly.

1.3 Compro

Compro is a plug-in that is developed and maintained by Convergent Manufacturing Technologies Inc. [1], which enables advanced composites process

TABLE 7.1 Composite material manufacturing process modeling software.

FiberSim SIEMENS	Composite fiber modeler DASSAULT SYSTEMES	Compro CONVERGENT MANUFACTURING TECHNOLOGIES · US	Aniform ANIFORM VIRTUAL FORMING
• Creates ply geometry by defining transitions with sequence, drop-off, and stagger profiles that automatically populate in the CAD model. • Simulates part producibility based on material and manufacturing process • Automatically generates manufacturing data such as flat patterns and information to drive automated cutting machines, laser projection systems	• Provides instantaneous flat patterns • Analyzes the producibility of a design based on theoretical projected fiber angles. Deviations show difference between actual and theory	• Calculates process induced deformations (e.g., spring-in) and develops recommendations for geometric compensation • Calculates residual stresses in composite laminates and configured structures • Iterate on cure cycles and tooling materials to ensure they meet the required specifications	• Mainly developed for thermoplastic material systems but can be modified to support thermoset materials • Material behavior and kinematics are tailored to large deformations • It can be used for producibility of manufacturing processes mainly compression modeling or stamping

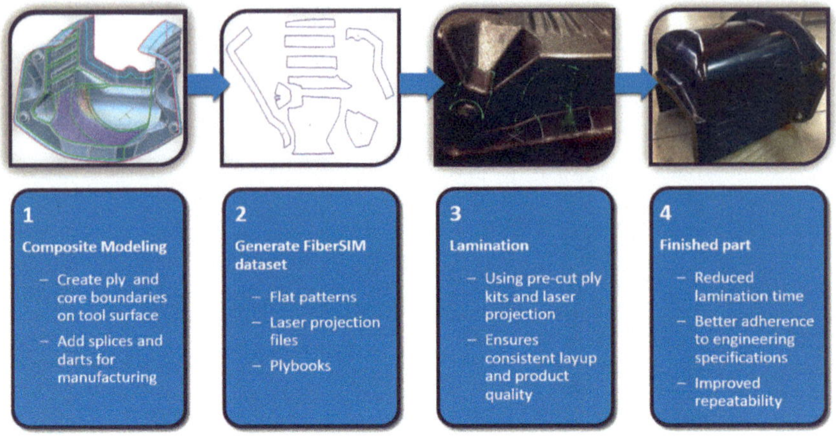

FIGURE 7.2 **FiberSim process flow from design to production implementation.**

simulation within general-purpose finite element environments such as ABAQUS and ANSYS. The tool is used for predicting spring-in calculations, tool compensation, cure modeling and much more. The premise relies on having a material system that is fully characterized by having its cure kinetics well understood that is compatible with the constitutive relations developed. For more information on the details the reader is referred to the dissertation by Johnston [2]. As far as this book is concerned, we want to offer the reader some guidelines on how to implement it in an actual part. The analysis starts by having a material system that is characterized (if it was not, it can be done working in collaboration with convergent) and a CAD model of the part/s that require analysis. This is followed by a finite element model of the part/s with mesh defined, element type, boundary conditions, and so on. The only difference comes when you assign the material which will be a material card offered by the plug-in compro. This is done for parts that we want to understand warpage, spring-back or just the heat-up rate during cure. A flow chart of this process is shown in Fig. 7.3. We also have a detailed example of how to conduct such analysis in Appendix that we recommend the reader follow to learn how such a model is setup, run and analyzed.

1.4 Aniform

AniForm is another software that is primarily used for thermoplastic forming prediction. Recently it has been looking at ways to incorporate the use of thermoset materials as well, but the primary focus is on thermoplastic and related processes. The software consists of a graphical user interface to define the process, and an implicit finite element solver to perform the calculations. Fig. 7.4 shows an outline of the process steps to run a model using the software.

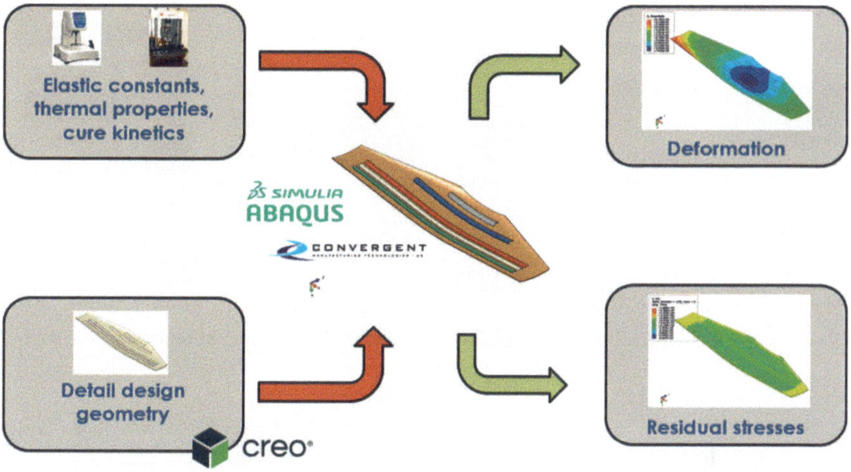

FIGURE 7.3 **Flow chart showing the inputs and outputs of a model that was analyzed via Compro/Abaqus.**

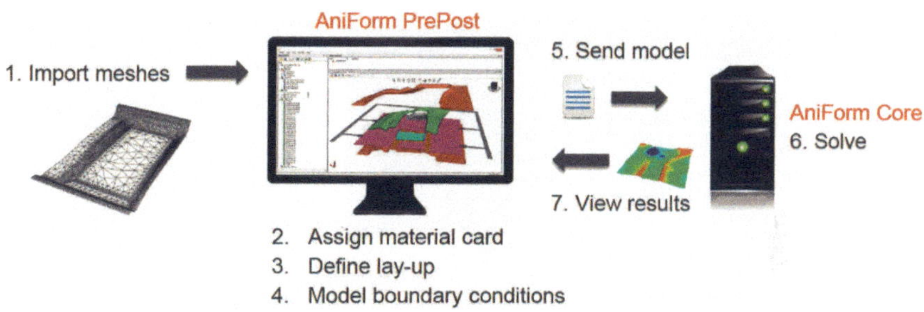

FIGURE 7.4 **Aniform software details and capabilities.**

The processes it supports are those used to build thermoplastic parts which are drape forming under pressure and temperature as shown in Fig. 7.5. These processes are known for their quick cure and part fabrication, which take approximately 30 min sometimes from positioning to final cure. People that have only used thermoset materials were intrigued by this process, which led the way to investigate approaches to consider such processes for their own fabrication. A recent summary was presented as SAMPE conference and some interesting results were shown that have potential for industry disruption in the next several years [3].

The results of the analysis provide the user with indications of possible areas of concern during the fabrication that should be considered more carefully to prevent quality issues. But unlike FiberSim which purely relies on geometric limitation inputted by the user, Anifrom considers the material behavior similar to how compro does but also considers the geometry so one can look at it as a

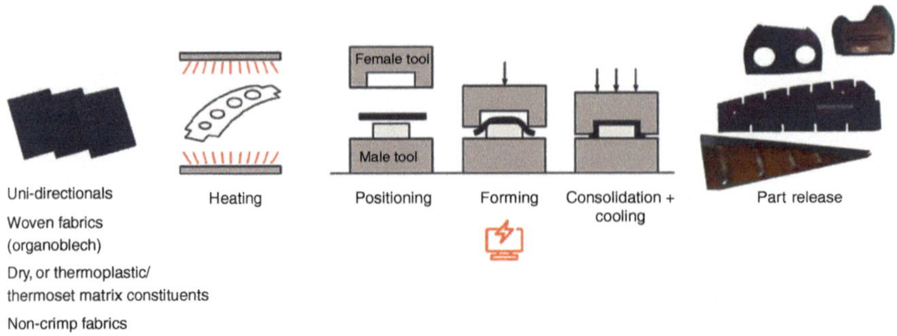

FIGURE 7.5 **Press forming process typically used in the fabrication of thermoplastic parts.**

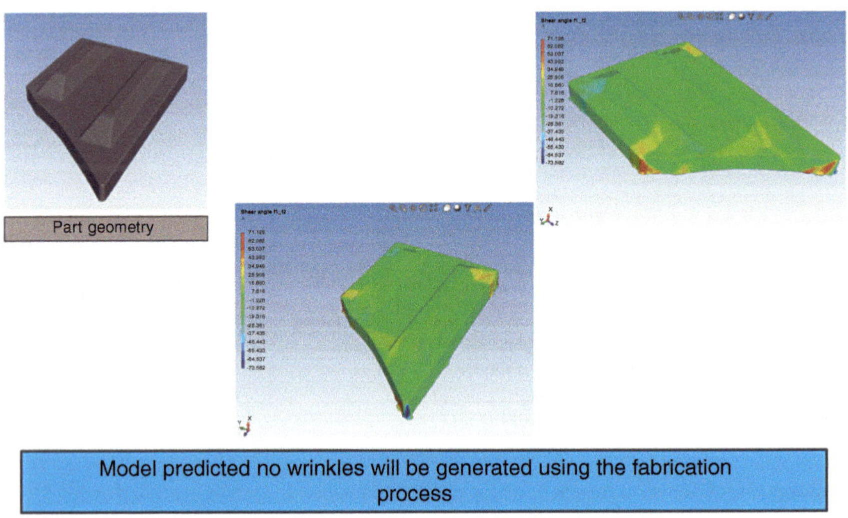

FIGURE 7.6 **Example of a part that was analyzed using Aniform and associated results showing no areas of concern.**

combination of both. The strain values can be set to some threshold to indicate when a wrinkle will get generated to inform the analyst on possible anomalies in the part during the analysis. An example of a part analysis is shown in Fig. 7.6.

1.5 VERICUT composite programming (VCP)

This software is primarily used when utilizing AFP or ATL and generates the data needed to communicate with the robot used to deposit the composite material as shown in Fig. 7.7. During the layout, the software can provide the user with information pertaining to areas of concern on the part based on the tool geometry and limitations of how the material can steer around. An example of

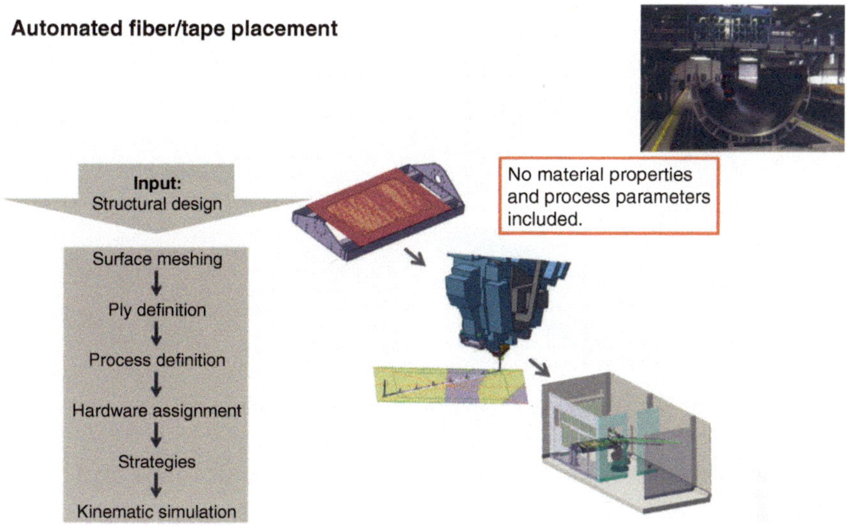

FIGURE 7.7 **Steps that VCP takes to implement during fabrication.**

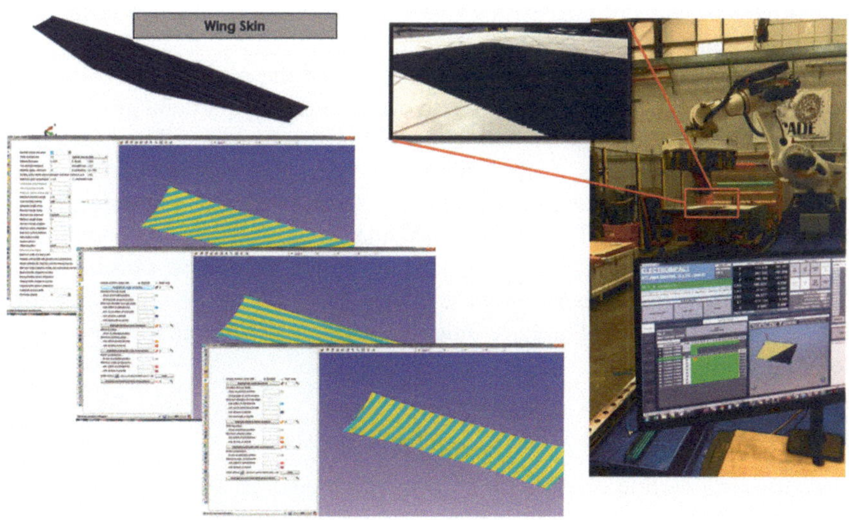

FIGURE 7.8 **Part analysis using VCP.**

such analysis is shown in Fig. 7.8 where a wing OML was used to analyze the ability to use 0.5-inch wide tape to laminate the part. All three lamina directions were looked at (0 degree, 45 degree, 90 degree) and for this particular example no issues were found.

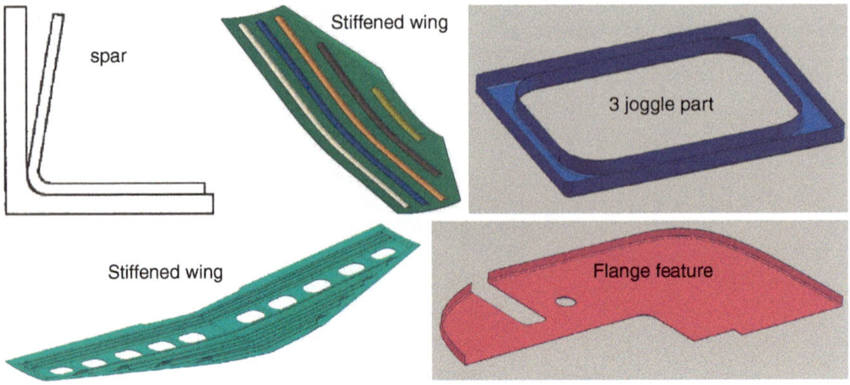

FIGURE 7.9 **Different type of parts that have unique design features.**

2 Modeling and simulation of composite structures and tools

After this brief introduction on the different software's available we will dive into some modeling and simulation as it involves tooling. In general, it is recommended to use many of the analysis tools available in order to mitigate risk downstream as many of the issues when it comes to tooling are very costly and time consuming to fix.

For warpage and spring-back which are two of the most prominent issues facing tooling, the recommendation is to evaluate part by part and provide recommendation on either compensating the tool or not. Note that if we were able to pull the part back to its shape without impacting the structural integrity and without the need to compensate the tool that is recommended. For spring-back, it is recommended to look at part families and specific design details (corners, flanges, etc.,) and provide design guidance on how to compensate the parts based on numerical and experimental analysis. Fig. 7.9 shows an outline of different part constructions that have unique features that will impact the level of deformation in each part.

2.1 Warpage analysis

The following section discusses the numerical modeling used in order to virtually predict the warpage of the wing skin component that was discussed in detail in Chapter 4. The details discussed in this section are not meant to be comprehensive rather a guide to inform the reader on the ability to perform such simulation using available commercial codes and supporting plug-ins. A correlation between the wing skin warpage measured and the simulation is shown that is used to validate the modeling approach and provide confidence in order to perform further evaluation in the future on different designs prior to fabricating parts.

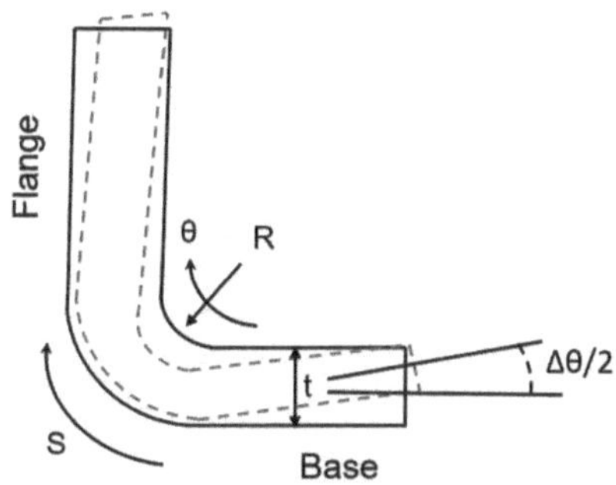

FIGURE 7.10 **A schematic representation of the spring-back phenomena in composite structures.**

When it comes to composite structures it is well known that having an anisotropic laminate creates thermal strains and residual stresses causing the part to warp. From that perspective, this phenomenon can be prevented by striving for a symmetric layup during the design phase. For curved sections however, differences in strains through-thickness, in the in-plane direction and geometry changes will result in a change to the part shape as shown in Fig. 7.10.

The first attempt to calculate the magnitude of the spring-back angle was proposed by Nelson and Cairns [4] using the following equation:

$$\Delta\theta = \theta\left[\frac{(\alpha_\theta - \alpha_R)\Delta T}{1+\alpha_R \Delta T}\right] \tag{7.1}$$

where θ is the angle between the base and flange, $\Delta\theta$ is the change in the angle between the pre-cure and post-cure state, α_θ is the laminate thermal expansion coefficient in the circumferential direction, α_R is the laminate thermal expansion coefficient in the radial direction and ΔT is the temperature change. This equation does not account for any through-thickness variations in temperature and resin degree of cure, which is acceptable when it comes to thin part, as those effects can be negligible. In thicker parts however, the low-composite transverse thermal conductivity coupled with the rapid-heat generation of the resin reaction may result in significant temperature and cure gradients through-thickness. Those gradients are potential sources of residual stresses in the part. In addition, if the boundary conditions during processing are not symmetric due to presence of process tooling, an asymmetric stress state may develop resulting in warpage. The uneven resin flow also results in uneven thermal and cure shrinkage strains. These strains result in non-uniform residual stress states and warpage in the part.

The tool also plays a role in the residual stress formation through the influence of temperature and mechanical loads and constraints applied at the tool part interface.

The impact of many of these variables have been studied by many researchers [2, 5–6] and they have shown the importance of understanding those details during the preliminary design phase in order to reduce their impact on the final product. However, there is lack in evaluating thick co-cured stiffened structures in the literature, which is one of the objectives of the current section, which is to implement well-known methods into such large-scale components.

When it comes to thermoset materials the stiffness of the resin significantly depends on the degree of cure and can be defined as [7]

$$E_r = (1-\alpha_{mod})E_r^0 + \alpha_{mod}E_r^\infty + \gamma\alpha_{mod}(1-\alpha_{mod})(E_r^\infty - E_r^\infty) \tag{7.2}$$

$$\alpha_{mod} = \frac{\alpha_{\theta-}\alpha_{mod}^{gel}}{\alpha_{mod}^{diff} - \alpha_{mod}^{gel}} \tag{7.3}$$

where E_r^0 and E_r^∞ are the fully uncured and fully cured resin moduli, respectively. α_{mod}^{gel} and α_{mod}^{diff} are the bounds on the degree of cure during the phase when the resin modulus is assumed to develop and γ is a parameter representing the competing mechanisms between stress relaxation and chemical hardening [7]. Generally, it is assumed that E_r^0 is equal to $E_r^\infty/1000$ as a first approximation [7].

When it comes to the geometry of the part and since we are dealing with stiffened structures the details of the design will have a large impact on the warpage as well. Assume that we select a representative element from the wing skin, and we divide it into beams based on the unique features it exhibits as shown in Fig. 7.11. Each beam will have its own behavior based on its characteristics but due to the need for continuity in traction and displacement at the interface [8] that will cause some asymmetry in the overall behavior which reflects on the warpage seen in the part.

And since we are dealing with thick structures the bending moment adheres to the following equation [9]

$$M = \frac{\Delta\theta}{\theta}ER_n(RA_m - A) = \frac{\Delta\theta}{\theta}E\frac{t}{ln\left(\frac{R_0}{R_i}\right)}\left(\left(R_i + \frac{t}{2}\right)bln\left(\frac{R_0}{R_i}\right) - bt\right) \tag{7.4}$$

where t is the beam thickness, E is the elastic modulus, b is the element width and R is the radius of curvature. In order to implement these variations, we will use the commercial FEM software Abaqus to simulate the cure behavior of the part to quantify the amount of warpage. In order to accurately model the

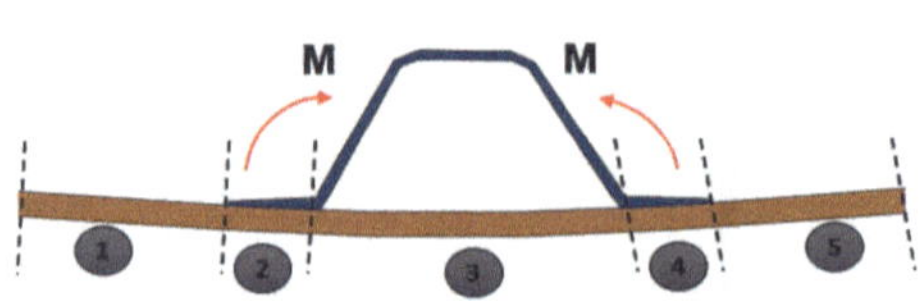

FIGURE 7.11 **Representative section of a stiffened hat structure.**

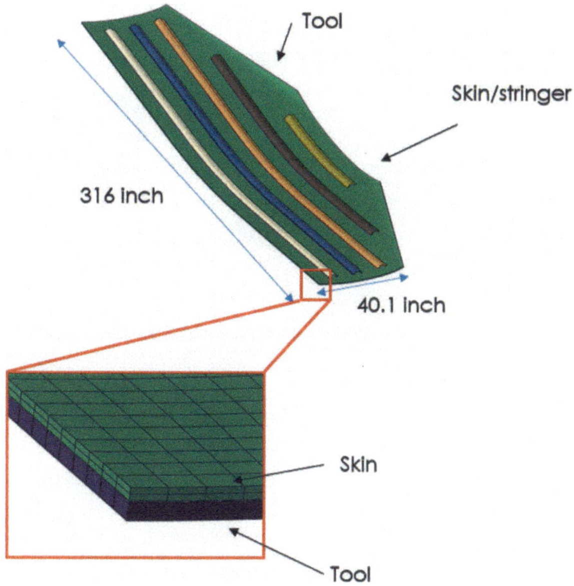

FIGURE 7.12 **Finite element model details.**

material behavior, the plug-in COMPRO will be utilized that uses a modified approach to define the material kinetics incorporating the temperature dependency as suggested in the cure hardening instantaneous linear elastic (CHILE) approach. Note that Cycom 5320-1 was fully characterized and available for use in this plug-in. The model used 8-node linear brick, incompatible mode C3D8I as the element type. The total degree of freedom (DOF) of the model was approximately 10 million with a runtime of 8 hrs using 8cpus on a supercomputer. The skin and hat stacking sequence mapped that of the actual design. An overview of the model is shown in Fig. 7.12.

In order to validate the model, we compared the warpage measurements taken from the wing demo once it was demolded while remaining on the tool as shown in Fig. 7.13. The results of the model are shown right above the measurements taken. We used a ruler to measure the displacement around the perimeter of the part and compared it to the global displacement of the FE model. The correlation among the points was within 33% at the corners and generally within 50% elsewhere. The deviation can be explained by the movement of the part during the de-bagging process as we do not account for that type of movement in the simulation. While this may not seem to be a great correlation result especially for those readers that are experienced with strain gauge correlations, for this type of thermo-mechanical simulation of the cure cycle it is a good indication in order to guide the engineers toward the correct path in the design and we are not necessarily interested in a perfect correlation. Overall, the simulation provided an accurate representation of the part behavior postcure.

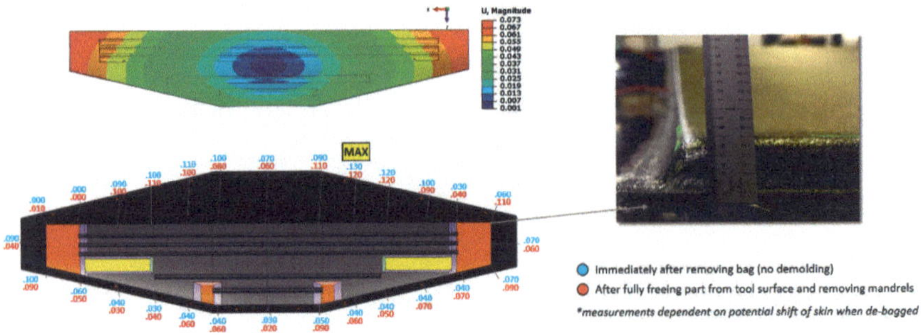

FIGURE 7.13 **Correlation between the model and the experimental results for the part prior to transferring it fully from the tool.**

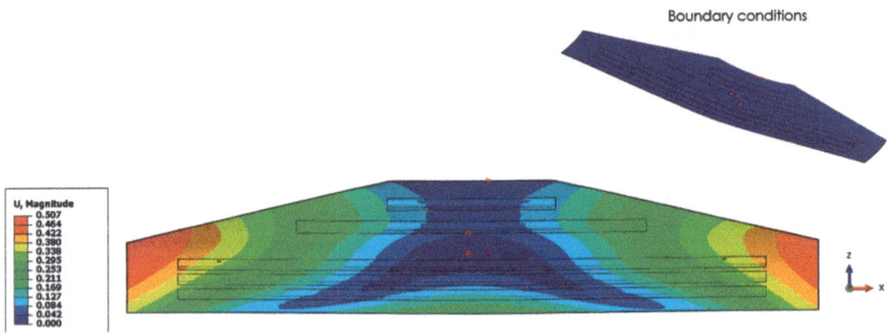

FIGURE 7.14 **Part displacement post-cure with no gravity effects on the part.**

We also simulated the warpage observed in the part in free deformation state assuming no gravity applied, and the results are shown in Fig. 7.14. The max computed displacement was 0.507 inches. The important question to answer was, at this displacement level, are we able to pull the part back into its nominal shape during assembly without causing major structural damage? We simulated pulling the part back into its nominal shape and extracted the resultant force needed to achieve that as shown in Fig. 7.15. It was determined that the load levels between 670 lbs and -2931lbs that are needed to pull the part back are not detrimental to the structural integrity.

Note that we did not compare the results from the experiment to the free deformation observed in the simulation. This was due to the inability to hold the part in a free deformation state at the time, but the study was mainly performed to evaluate the ability to pull the part back while the part was on the tool and demolded. The final part held at three boundary locations is shown in Fig. 7.16.

This demo article provided insight to the flexibility of co-cure structures compared to a cobonded structure where the later require additional forces to pull the part back to nominal shape causing the need to consider warpage more

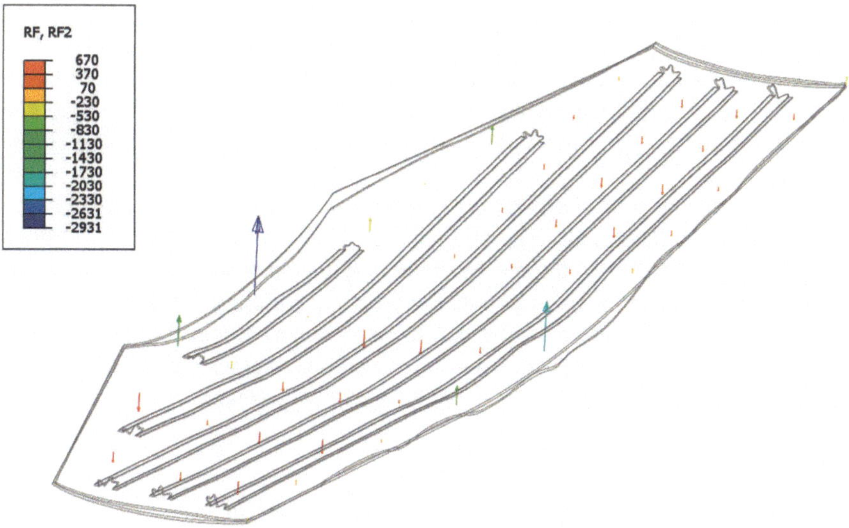

FIGURE 7.15 **Reaction forces needed to pull the part back to nominal shape.**

FIGURE 7.16 **Final wing skin part.**

closely (this was based on experience conducting work on co-bonded wing skins for the Boeing 787 and Boeing 777X).

The hypothesis behind why co-curing offers more flexibility compared to co-bonding is that when you co-bond, there is usually an uncured part attached to another cured part. During the second cure process to make both parts into a single part (e.g., uncured wing skin bonded to cured stringers), there are larger reaction forces caused by the cured stringers onto the uncured skin which translate into added residual stresses in the part that are released once the part is demolded creating large amount of warpage. While in the co-curing process, there is only a single cure and the material state allows the parts to deform simultaneously in order to reach the equilibrium state with less forces being generated at the interface that translate into less warpage overall.

2.2 Residual stresses

In this section, we evaluate the amount of residual stresses generated in a co-cured hat-stiffened composite structure using a high-definition fiber-optic sensing (HD-FOS) technology embedded within the panel during fabrication. The results will shed light on the amount of residual stresses that are expected to be generated in this type of structure that is typically used on aircraft wings and fuselage components. The results can be used to tune the structural analysis methods by potentiality applying a knockdown to the allowables as required in order to account for the impact of the residual stresses on the structural integrity of the part.

An out-of-autoclave material system namely Cycom 5320-1 resin will be used to fabricate the panels. The UD material form will have IM7 as the fiber of choice while the PW will have T650 fibers. This study will also compare the results obtained from the experiments to numerical simulation done via compro and abaqus to offer a method that can be used during the preliminary design phase to account for the expected amount of residual stresses virtually. This will allow rapid design changes and prevent escapes downstream.

2.2.1 Experimental evaluation

In order to appropriately understand the behavior of the co-cured stiffened structure that was being considered as the primary design to build the wing skins from, the potential residual stresses generated during the cure process needed to be quantified and the associated warpage. The reason for the concern was the large size of the wing skins and thickness, and any impact had the potential to adversely impact the structural integrity. We approached the evaluation by following a building block methodology as shown in Fig. 7.17. We started with

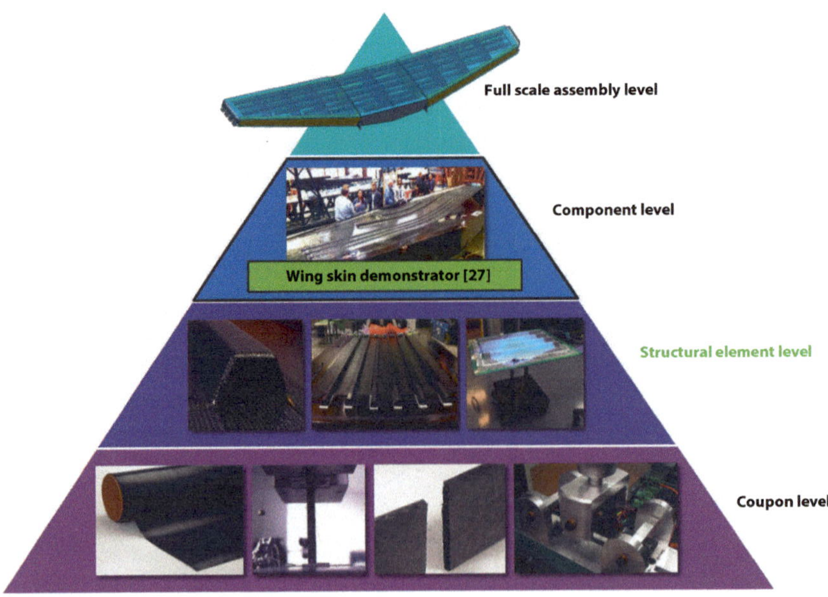

FIGURE 7.17 **Manufacturing building block approach established for the program.**

smaller-scale material testing to understand the material physical behavior and allowables. This was followed by element level testing which is the focus of this section. Additional work was done on the component level and the reader is referred to reference [10] for more information on those details.

2.2.2 Fabrication

Two panels were built for this study, the first panel was cured in an oven and used a rolled tow-filler design. The primary objective was to evaluate the residual stresses in the panel using a HD-FOS system offered by LUNA, Inc that were embedded in the panel during fabrication. The second panel was cured in an autoclave using a preformed tow-filler design. The objective in this case was to evaluate the warpage seen in the panel and correlate it back to the analysis. Nondestructive and destructive inspections were used to quantify any manufacturing process defects and panel quality post fabrication. An outline of the panel is shown in Fig. 7.18.

One of the main manufacturing variables that will impact the quality of the part build is the tow filler design that is used to fill the gap created during the lamination of the hat stiffeners as shown in Fig. 7.19. It is well known that the filler is considered a critical structural detail of the part that can have an adverse impact on the structural integrity. Therefore, additional emphasis needs to be

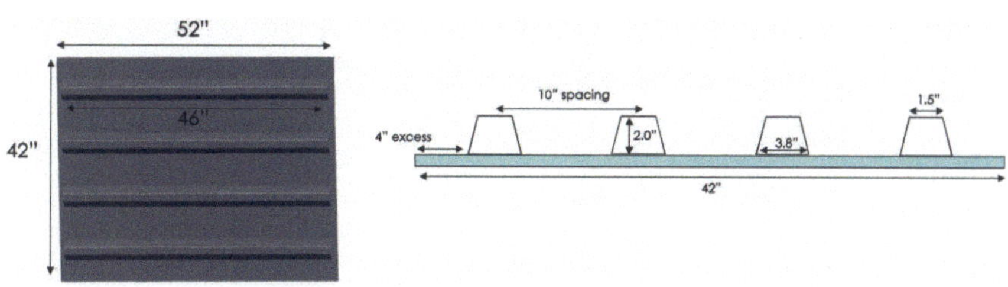

FIGURE 7.18 **Composite hat stiffened panel outline.**

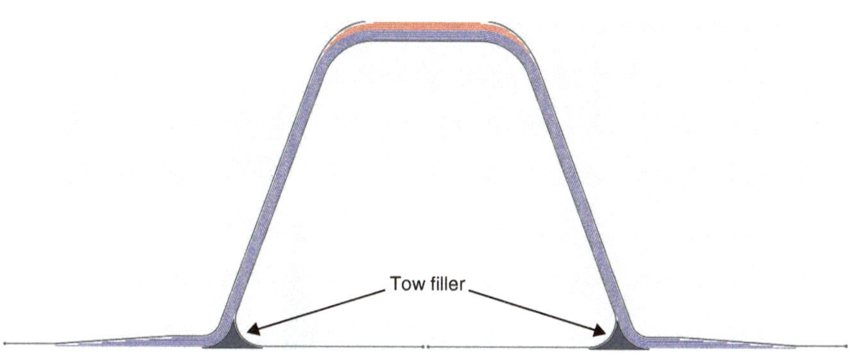

FIGURE 7.19 **Tow fillers used during the fabrication of composite hat stiffeners.**

given to the fabrication process of this feature in order to reduce any anomalies that can impact its quality.

The rolled tow filler fabrication approach is shown in Fig. 7.20 where rolled UD tows were made from 7" by 46" strips of IM7/5320-1 tape. Those rolled fillers were then located near the hat mandrel web using a special tool and heat gun to form in place. The preformed tow filler on the other hand utilized a tool that was specially designed based on the actual size and volume of the hat section. This created tow filler that fitted perfectly around the mandrel web once complete as shown in Fig. 7.21.

The mandrel design is the other important variable to consider during the fabrication of hat section stiffeners. In this study, an extruded mandrel was used for the fabrication of the panels as shown in Fig. 7.22. Depending on the expansion during cure this can have an adverse impact on the final part quality. For this type of mandrel, a 5% expansion was expected and the design was adjusted to account for this change.

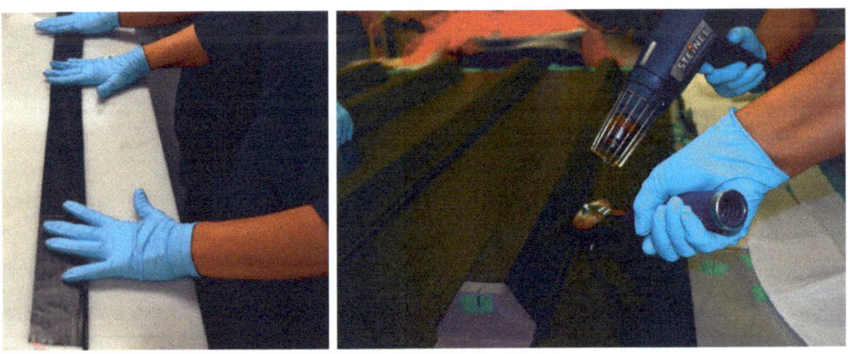

FIGURE 7.20 **Rolled tow filler fabrication approach.**

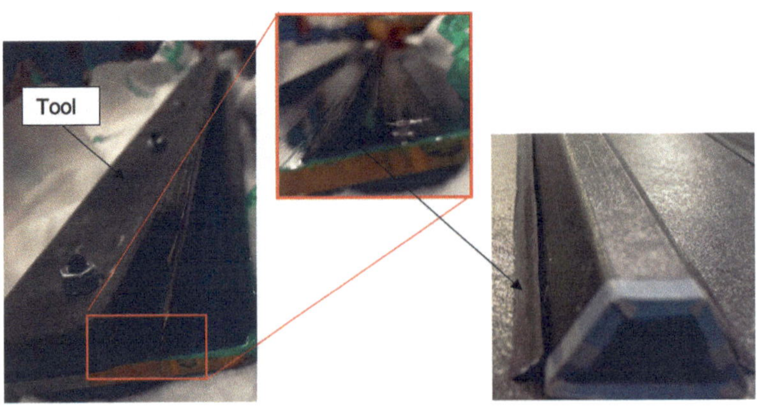

FIGURE 7.21 **Preformed tow filler fabrication approach.**

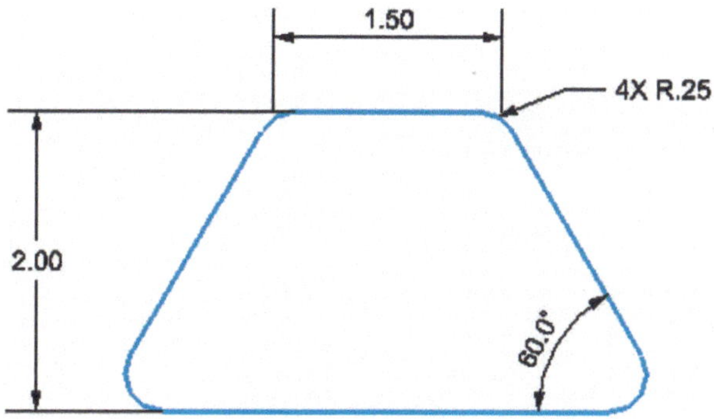

FIGURE 7.22 **Mandrel design used for the hat stiffener fabrication.**

Hand-layup was used as the primary method of lamination since it was considered our baseline process for building the part during production. This study gave the development team an opportunity to capture any critical details that need to be adjusted in the process specification prior to building the production parts. The skin included 63 plies of UD material form and 2 plies of PW form located on the OML and IML of the part. The hat stiffeners were made from 18 plies on the web and an additional 15 plies of pad-up near the cap section as shown in Fig. 7.23. The mandrel was wrapped with 1 ply of PW and a final PW ply as a closeout to the entire panel. The details of the skin layup, locating the stiffeners onto the skin and the final panel part prior to bagging and cure is shown in Fig. 7.24.

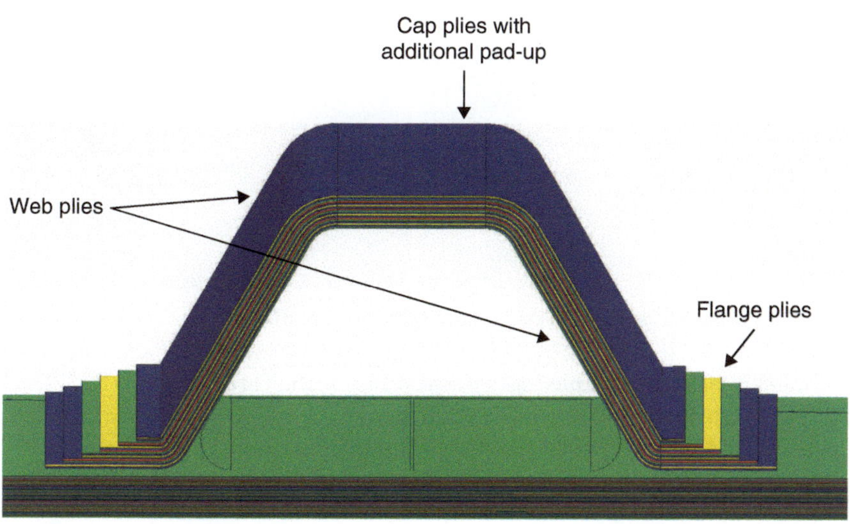

FIGURE 7.23 **Hat stiffener design.**

FIGURE 7.24 **Fabrication steps of the hat stiffened panel.**

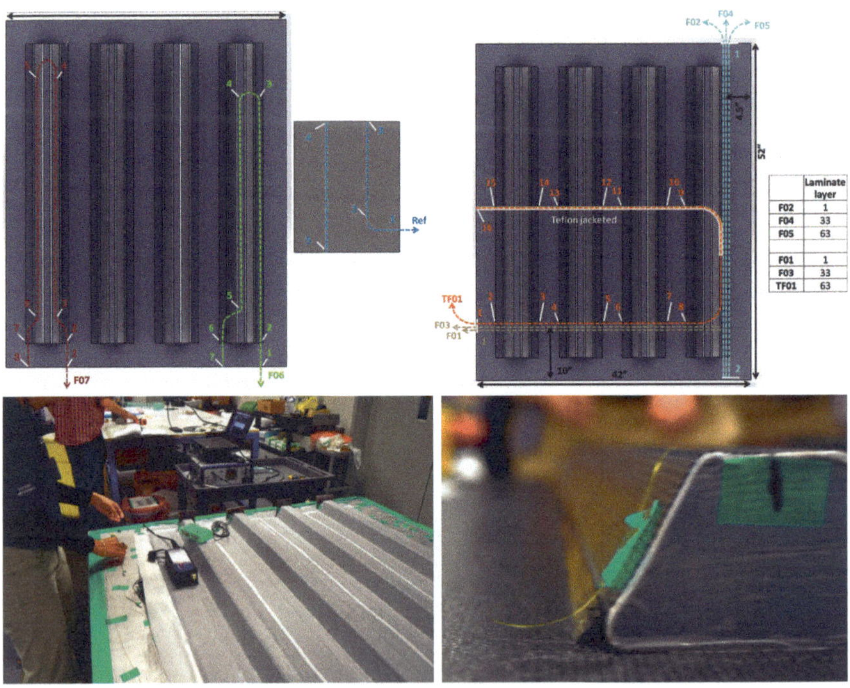

FIGURE 7.25 **Fiber optic strain sensor installation and layout.**

Panel 1 was instrumented as shown in Fig. 7.25. Three sensors (F02, F04, F05) were laid longitudinally under the flange of a hat stiffener, at 3 layers within the laminate (1, 33, 63 with layer count starting at the tool), entering and exiting the panel. Three sensors (F01, F03, TF01) were laid in the transverse direction, at the same 3 layers within the laminate, with the sensor termination ending just before its matching perpendicular sensor. Sensor TF01 made a U-turn to exit the

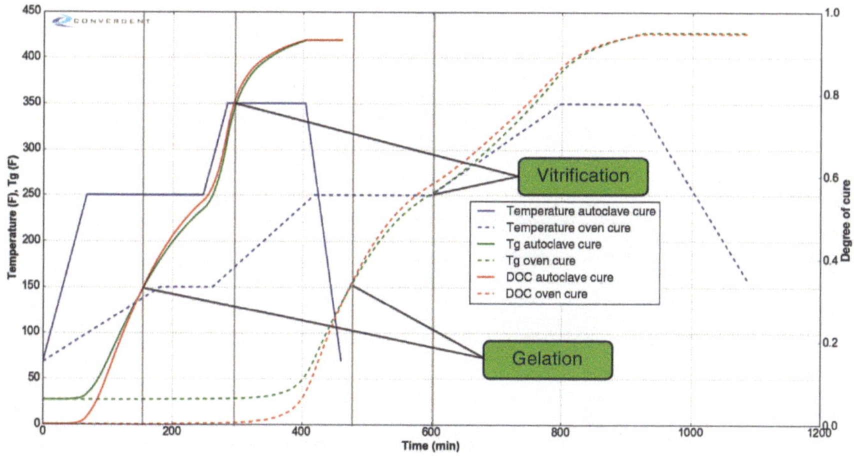

FIGURE 7.26 **Cure cycle comparison between oven and autoclave cures.**

panel at the midline. The last ~42" of this sensor was housed in a Teflon jacket, in order to isolate it from strain, therefore reporting only temperature change along its length. Sensor TF01 was continuously monitored throughout the curing process. Sensor F06 was placed within a hat stiffener, where it traversed the length of the hat flange and made its way back down the top of the hat. Sensor F07 was also placed within a hat stiffener, where it traversed the two slopes of the hat. All strain measurements were zero-ed (Tared) just before bagging. A total of 12 thermocouples were also installed within both panels.

The cure cycles for the oven and autoclave cures are shown in Fig. 7.26. RAVEN process simulation software [11] was used to predict the development of DOC and the glass transition temperature (Tg) for our material system that was fully characterized in the software. Based on this result, the gelation and vitrification points are identified in the same figure. The gelation in the material occurs around the DOC of 0.30 [12], and vitrification occurs when Tg ~> T_{cure}. Moreover, note that the oven cure is almost twice as long as the autoclave cure to ensure that the appropriate level of cure is achieved with the lack of pressure. Panel 2 that was cured in the autoclave did not include any fiber-optic sensors; this was done in order to isolate the comparison between both panels.

2.2.3 Results

Residual strain measurements postcure, after unbagging, and after popping off the tool, were thermally compensated using the average thermocouple measurement from all 12 thermocouples. The following steps were taken for thermal compensation:

- Convert temperature from °F to °C
- Calculate temperature change in °C

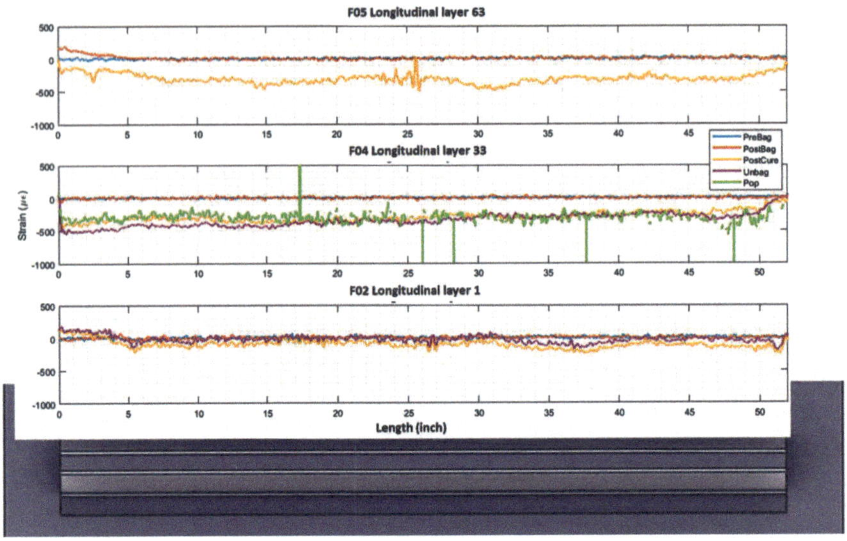

FIGURE 7.27 **Residual strain measurements from the longitudinal stack of sensors before bagging, after bagging, after curing, after unbagging, and after popping the panel off the tool.**

- Convert this temperature change to equivalent strain by dividing the temperature coefficient and multiplying by the strain coefficient
- Subtract 87% of this value from the measured strains (87% of a fiber sensor's thermal response is due to changes in index of refraction)

The temperature spread between thermocouples after cure was approximately 10 °F. By the calculations above, this results in a strain compensation spread of 40 με. In other words, the uncertainty associated with this compensation method attributed to the variation in thermocouple readings is +/−20 με.

Strain measurements from the longitudinal stack of sensors located within the panel, under the hat flange, are shown in Fig. 7.27. There is typically no strain change due to the bagging process. The sensor closest to the tool face experiences very little residual strain from the manufacturing process. However, the sensor in the mid-plane of the panel (F04) and the sensor close to the top surface of the panel (F05) experience roughly 500 με of compression fairly uniformly along their length. The reflections seen in sensor F04 is due to the sensor breaking close to the termination when the panel shifted on the tool as a silicone gasket was aggressively pulled out. This break created a large back-reflection, increasing the noise floor. Therefore the final measurement of strain once the panel was popped off the tool (green trace) has intermittent gaps in data as well as large measurement excursions.

Sensor F05 displays some ripple along its length after the cure cycle, indicating periodic non-uniformity at that particular layer. Its strain profile also includes an area of sharply changing strain gradients right in the middle of the

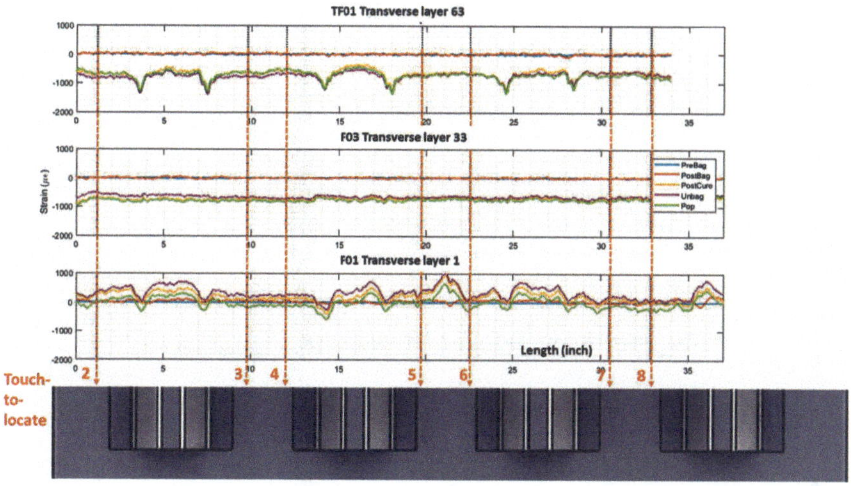

FIGURE 7.28 **Residual strain measurements from the transverse stack of sensors before bagging, after bagging, after curing, after unbagging, and after popping the panel off the tool.**

sensor length. This sensor was accidentally cut during the unbagging process, preventing the measurement of its final strain profile.

Strain measurements from the transverse stack of sensors located within the panel, 10" from the edge, are plotted in Fig. 7.28. Touch-to-locate [13] was carried out at the flange edges of each hat, and are marked here by dashed orange lines. Similar to the longitudinal stack, there is no strain change due to the bagging process. Also similar to the longitudinal stack, the sensor closest to the tool experiences strains that average around 0 με after cure. However, there are obvious negative strain peaks that seem to correspond to the location where the hat stiffener terminates near the flange chord-wise. Sensor F03 in the mid-plane of the panel experiences roughly 700 με of compression fairly uniformly along its length. Sensor TF01 experiences a similar level of uniform strain along its length, with obvious negative strain peaks that once again correspond to the location where the hat stiffener terminates near the flange chord-wise.

Strain measurements from the two sensors installed on various surfaces of the hat stiffener are shown in Fig. 7.29. Once again there is no strain change due to the bagging process. For the sensor lengths along each web (F07), the residual strain post-cure is very uniform at -200 με. However, once the part is unbagged and the silicone gasket removed, the residual strain doubles in compression to approximately −400 με. Additionally, the strain profile becomes more uneven most likely due to the removal of the silicone gasket. The hat top (F06) is also in a compressive state post-cure, though there is minimal difference in strain states between post-cure and unbagging, while the hat flange (F06) sees some compressive strain being relaxed once unbagged.

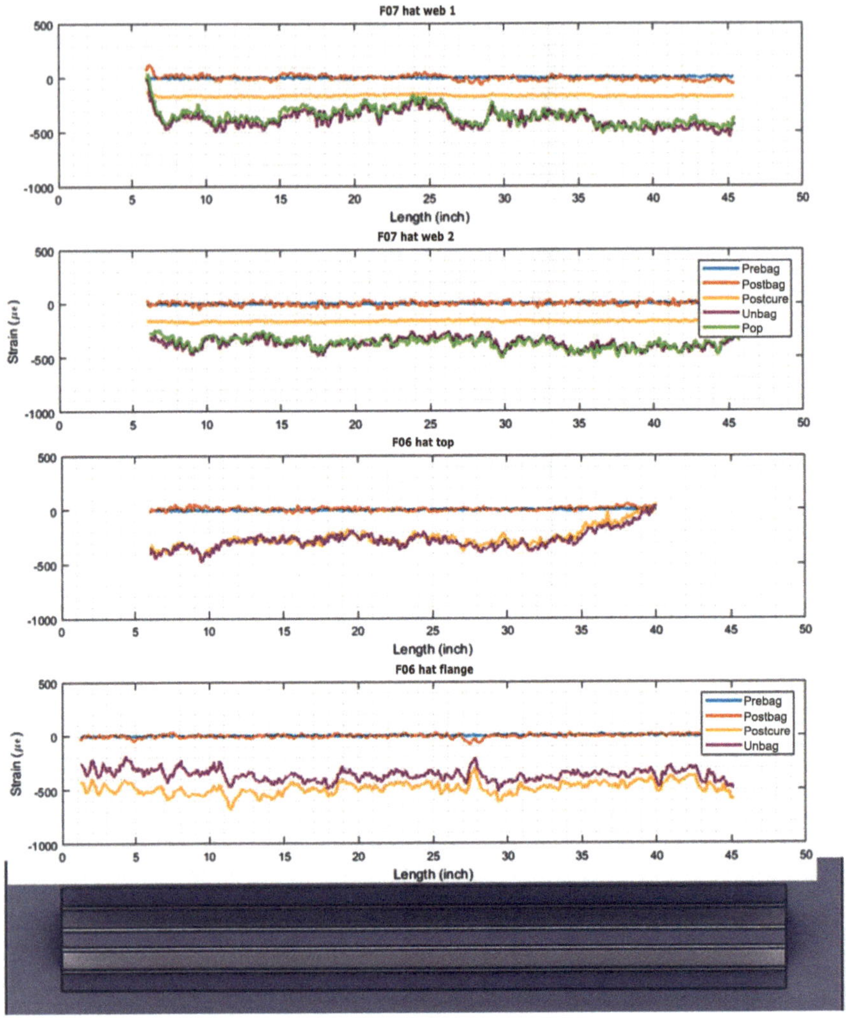

FIGURE 7.29 **Residual strain measurements from the sensors installed on various surfaces of the hat stiffener before bagging, after bagging, after curing, after unbagging, and after popping the panel off the tool.**

Thermocouple TC09 was co-located in the middle of the panel near the Teflon-jacketed section of sensor TF01, which was reporting temperature. A comparison between TC09 readings and the fiber-sensor measurement is plotted in Fig. 7.30. This shows very close agreement, with the fiber sensor over-reporting at the top of the cure cycle by ~10 °F. The uniformity of the temperature profile along the sensor is represented by the error bars along the fiber temperature measurements which average ±5 °F.

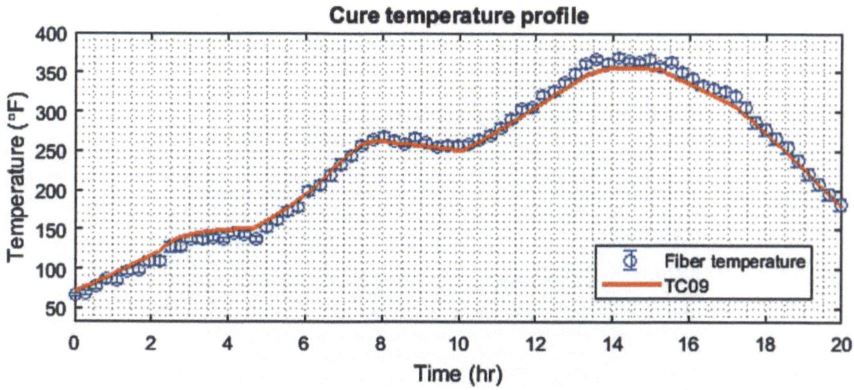

FIGURE 7.30 Temperature profile comparison between thermocouples and fiber optic temperature sensor.

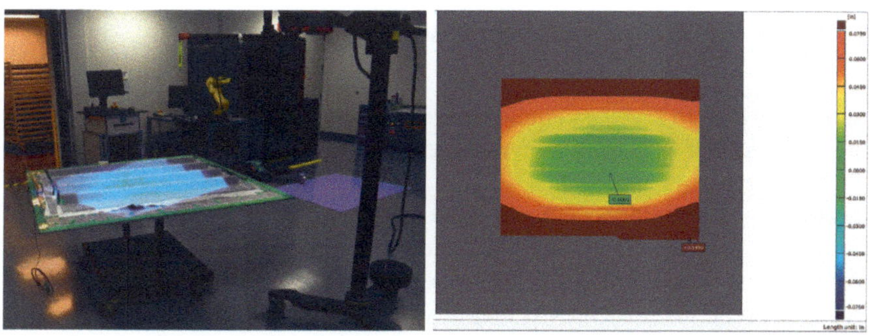

FIGURE 7.31 Blue light scan used to measure the part deformation post-cure.

Now in order to quantify the amount of warpage in the hat panel, a pre-cure and post-cure scan of the panel was taken using blue light technology [14] and analyzed. The results of the scan is shown in Fig. 7.31 where it shows the OML surface displacement. As observed, the maximum amount of displacement was seen near the panel edges with a maximum displacement magnitude of 0.149 inches. This behavior correlated well with how these types of structures deform in practice. It is particularly important to ensure that the level of warpage is minimal or manageable by using finite amount of force or vacuum to pull the part back to nominal shape during assembly without impacting the structural integrity of the part to ensure no fit-up issues will occur.

The reason for the warpage can be attributed to some asymmetry in the layup and geometry variations. Even though a symmetric and balanced layup was used for the skin, the stiffener included ramps near the cap region and termination that were not fully symmetric. In addition, the nature of this geometry where you have a stiffener attached to a skin and gaps between the different stiffeners can cause changes in the displacement across the panel. With

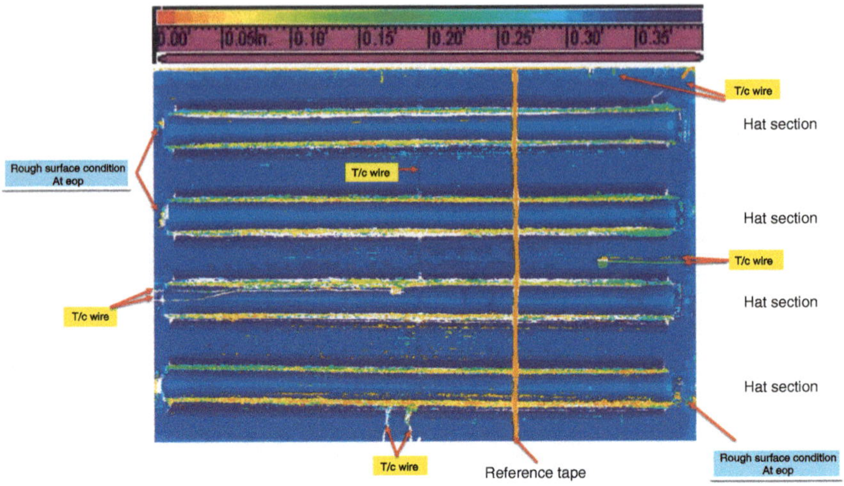

FIGURE 7.32 **Nondestructive inspection of hat stiffened panel.**

the panel trying to reach an equilibrium state, this causes some regions to have larger curvature component that translates into the warpage we observe.

Another part of the evaluation was the nondestructive and destructive inspections. A C-scan of the part was done in order to identify the part quality and ensure it is free of defects. Fig. 7.32 shows the results of the scan. The fringe plot shows the time of flight (TOF) reflection of the signal as it bounced back during the scan. As you can see, majority of the panel falls in the blue color domain of 0.36 inches, which corresponds to the skin thickness confirming that the scan reflected across the entire thickens with no detectible defects. However, it is clear that the tow filler region is not detectable using the C-scan and requires further evaluation.

The two panels were section cut and polished for further microscopic imaging analysis specifically near the tow filler region as shown in Fig. 7.33. The images clearly show voids in the panel that was oven cured and included the rolled tow fillers while the panel that was autoclave cured using the preformed tow did not show any voids or defects. This might be attributed to the size of the rolled tow filler during fabrication where it was too small causing consolidation issues during cure entrapping air in the laminate. For any future build, it is recommended to use preformed tow filler based on the volumetric calculation to eliminate this issue from rising.

2.2.4 Analysis and simulation

The experimental work shown thus far provided a great deal of insight into the behavior of hat-stiffened panel behavior that can be leveraged to design full scale parts for the platform and support the build process when it comes

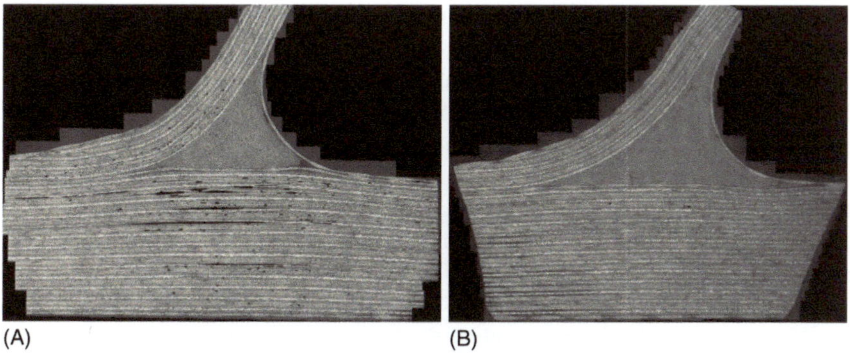

FIGURE 7.33 **Tow filler microscopic evaluation for (A) oven cure with rolled tow filler and (B) preformed tow filler autoclave cured.**

to production. However, the experimental determination of residual stresses in prototype parts can be exorbitant, both in terms of financial and temporal costs. As an alternative to physical measurement, it is possible for computational tools to be used to quantify potential residual stresses in composite parts as well as the warpage behavior. Therefore, this section focuses on the development of a simplistic approach for simulating the residual stresses and warpage using available commercial software and use the experimental data collected for validation.

Before introducing the modeling approach we want to expand on the variables impacting the generation of the residual stresses. It is known that the expansion coefficient of polymer matrix materials is usually much higher than that of the fibers and the expansion of the fibers is orthotropic in nature with very low or slightly negative expansion coefficients in the fiber direction, but higher values in the transverse direction for carbon fibers. This leads to residual stresses being generated at the microscale during cool down with compressive stresses generated along the fibers, with tension in the matrix in the fiber direction. The difference in ply-level expansion coefficients in the fiber and transverse directions is another mechanism that causes in-plane stresses in laminates, which can be analyzed by classical lamination theory. These can lead to distortion in flat plates when lay-ups are not balanced and symmetric. The high through-thickness expansion coefficients (matrix dominated) compared with in-plane values (fiber dominated) is another large variable impacting the generation of residual stresses. This causes a change in curvature within laminates with temperature for any lay-up, and is the origin of spring back phenomenon in composite parts.

The shrinkage in the matrix material during the cure produces an additional volume change. This can be a very substantial effect with 5% volumetric shrinkage during cross linking typically seen for thermoset materials [15] which can cause in-plane stresses and distortions in laminates, and changes in curvature in curved plates.

One final reason that can cause stresses in the part is the tool part interaction. Stresses can arise during the cure due to differential strains between the part and tool on which it is manufactured. Aluminum or steel tools have much higher expansion coefficients than composite parts, and tend to stretch the parts as they heat up. Gradients of in-plane stresses through the thickness are generated, causing bending when the stresses are released. A second tool–part interaction mechanism is due to locking, where the geometry of the part forces it to move with the tool as it expands. The simulation approach shall be able to incorporate these variables in order to accurately mimic the fabrication methods proposed

When it comes to the constitutive modeling relation that governs the cure-dependent instantaneous stress-strain relation of the thermoset composites during curing, many researches have proposed different methods [16–20].

Eq. (7.2) showed the resin modulus as a function of DOC, while the instantaneous resin shear modulus is given as:

$$G_m = \frac{E_m}{2(1+\nu_m)} \tag{7.5}$$

where ν_m is the Poisson's ratio of the resin and assumed constant due to the fact that the largest deviations in Poisson's ratio occurs at very low degree of cures, that is, where the material modulus is still undeveloped. In general, E_m and G_m are a function of time and degree of cure, but only indirectly a function of temperature through the degree of cure. In order to incorporate temperature dependency, the relationship between the resin degree of cure and the glass transition temperature Tg can be considered. The main effect of the included temperature dependency is the ability to model softening at elevated temperatures. The proposed model follows [2]:

$$E'_m = \left\{ \begin{array}{lc} E^o_m & T^* < T^*_{C1} \\ E^o_m + \frac{T^* - T^*_{C1}}{T^*_{C2} - T^*_{C1}}\left(E^\infty_m - E^o_m\right) \quad for & T^*_{C1} \le T^* \le T^*_{C2} \\ E^\infty_m & T^* > T^*_{C2} \end{array} \right\} \tag{7.6}$$

$$E_m = E'_m\left[1 - \alpha_{Er}(T - T_0)\right] \tag{7.7}$$

where T_{C1} and T_{C2} are the critical temperatures marking the bounds determining the linear variation of the modulus with T*, defined as:

$$T^* = \left(T^0_g + \alpha_{Tg} \cdot \alpha\right) - T \; and \; T^*_{C1} = T^*_{C1a} + T^*_{C1b} \cdot T \tag{7.8}$$

In Eq. (7.8), the temperature T* can also simply be modeled as T* = Tg − T. The remaining constants in the equations above are fitting parameters used to capture experimental data.

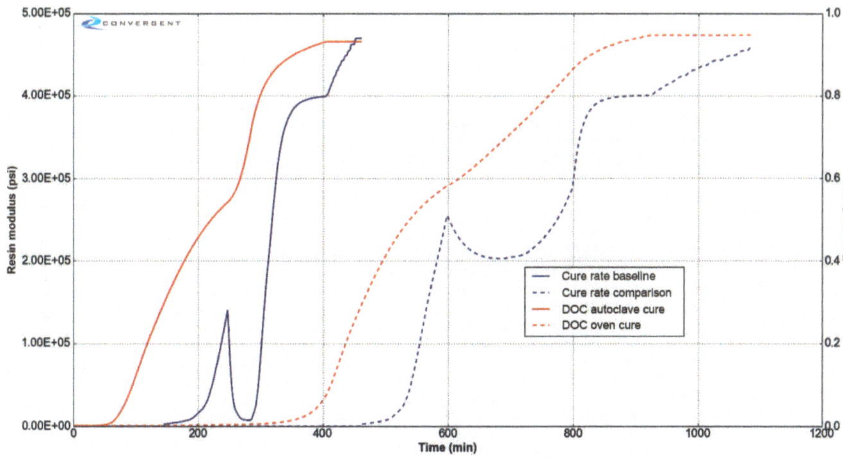

FIGURE 7.34 Example of the CHILE matrix modulus development as a function of the cure degree.

The approach above has been named the CHILE resin model. This designation indicates that the modulus of the instantaneous linear elastic resin increases monotonically with the progression of cure [20]. An example of the cure dependent resin modulus development when using the CHILE approach is presented in Fig. 7.34.

The instantaneous resin mechanical properties calculated are used to update the effective composite material properties using the self-consistent field micromechanics (SCFM) micromechanics approach. Following this, stress increments are determined as:

$$\Delta\sigma_{ij} = C_{ijkl}(\alpha, T)\Delta\varepsilon_{kl} \tag{7.9}$$

using the cure and temperature dependent stiffness matrix C and the elastic strain increment. The new stress at time $t + \Delta t$ is updated following:

$$\sigma_{ij}(t+\Delta t) = \sigma_{ij}(t) + \Delta\sigma_{ij} \tag{7.10}$$

where t is the current time and Δt is the time increment. The CHILE approach was incorporated in a plug-in called COMPRO as mentioned earlier that was developed to be used in conjunction with the commercial FEM software Abaqus to simulate the cure behavior of composite parts. Note that Cycom 5320-1 was fully characterized and available for use in this plug-in. We will be using both commercial codes for our analysis in this case. The model used an 8-node linear brick, incompatible mode C3D8I as the element type. The total degree of freedom (DOF) of the model was approximately 3 million with a runtime of 8 hrs using 8cpus on a supercomputer. The skin and hat-stacking sequence mapped that of the actual design. An overview of the model is shown in Fig. 7.35.

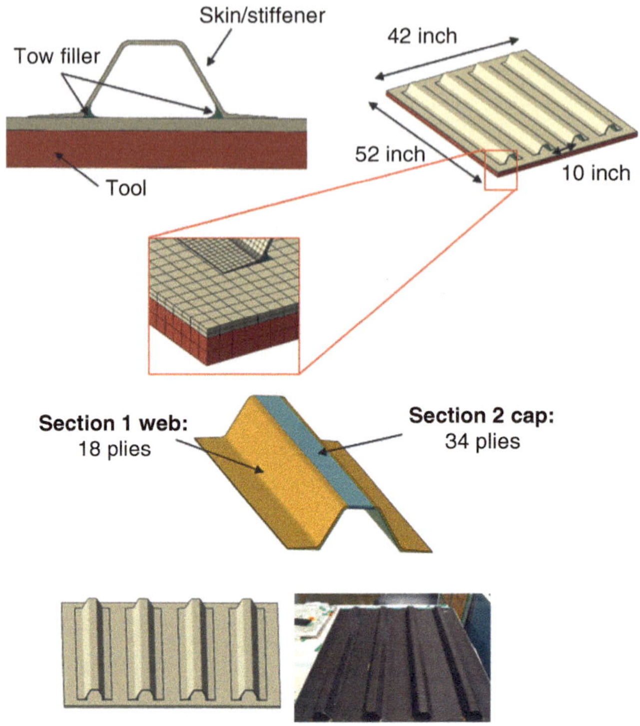

FIGURE 7.35 **Finite element idealization.**

A comparison between the strain values extracted from the model and the experimental results is shown next. Fig. 7.36 shows the longitudinal strain value comparison near the skin where the fiber optic sensor was embedded in the panel at layer 33 which is the middle of the laminate. The comparison was made at the step where the part was removed from the tool (popped off the tool). As shown, the simulation overestimates the value of the residual strains in the longitudinal direction by an average of 19%. With modification to the mesh density it might be possible to reduce the difference slightly but the runtime will increase and the analyst needs to make a decision on the level of accuracy needed based on the application. With approximately 500 $\mu\varepsilon$ this can be considered a high value and will need to be addressed during the structural analysis evaluation as the allowable used to compute margins do not account for this level of residual strains.

A comparison between the transverse residual strains and the simulation is shown in Fig. 7.37. Compared to the longitudinal strains this comparison is much more accurate within 5% in most locations. The dips in the strains levels correlate to the hat termination areas as you go along the chord direction which is also being captured clearly via the simulation. The small shift is due to the mesh density and having a more refined mesh could provide an even clearer

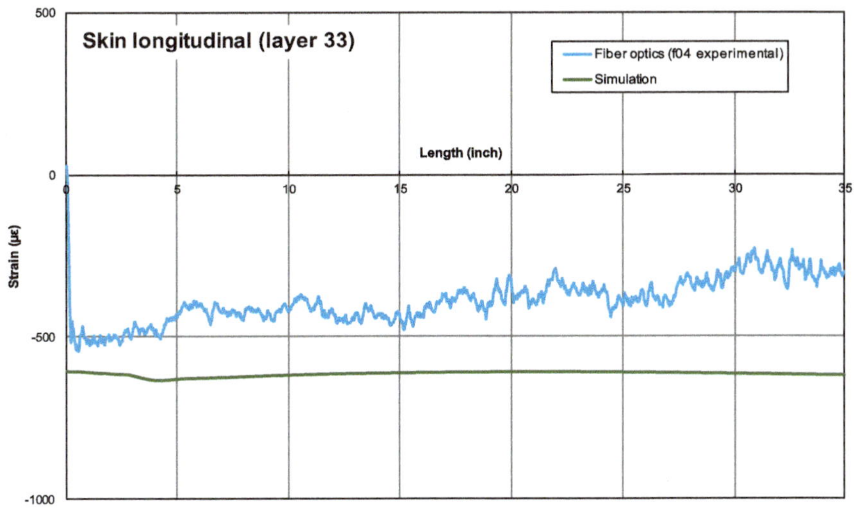

FIGURE 7.36 **Longitudinal residual strain comparison.**

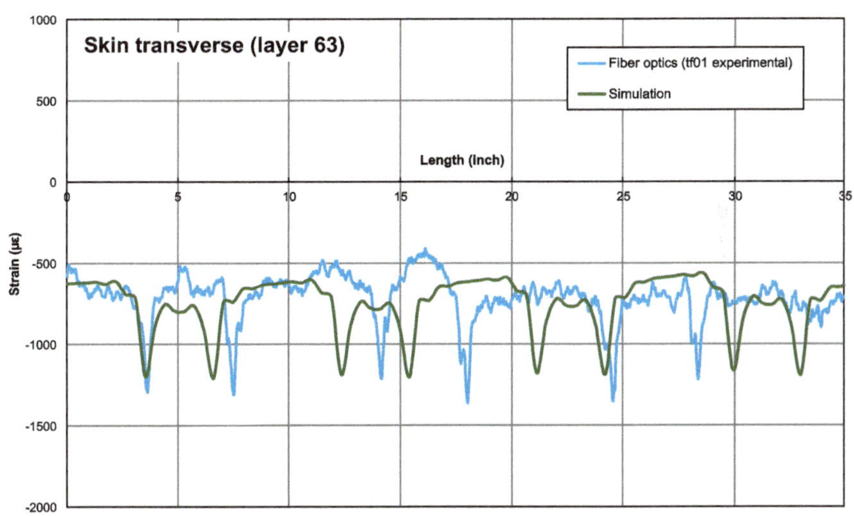

FIGURE 7.37 **Transverse residual strain comparison.**

alignment in location. The transverse residual strains are also higher than that of the longitudinal strains especially near the hat termination which emphasizes the importance of inspecting those regions during the production phase as they can cause premature failures if close attention was not given during the design phase. This also correlates well with the observations made that the tow filler region is a large area of concern when it comes to hat-stiffened structures and the hypothesis that residual strains can be the cause of the cracking seen is valid based on the strain values seen which are very high. This in combination with manufacturing anomalies can be a recipe for failure. Emphasis on ensuring a manufacturing process that minimizes any defects is crucial in those regions.

A comparison of the residual strains seen in the web and flange sections of the hat is shown in Figs. 7.38–7.39, respectively. The strain values from the

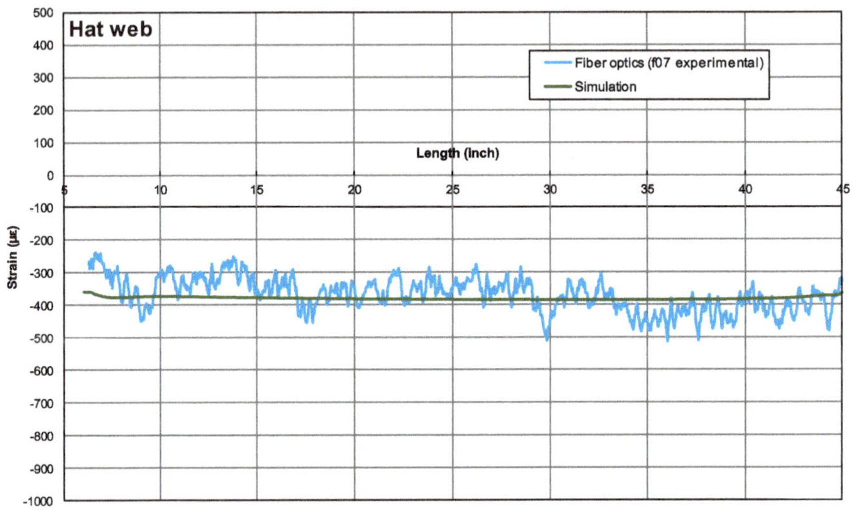

FIGURE 7.38 **Hat web residual strain comparison.**

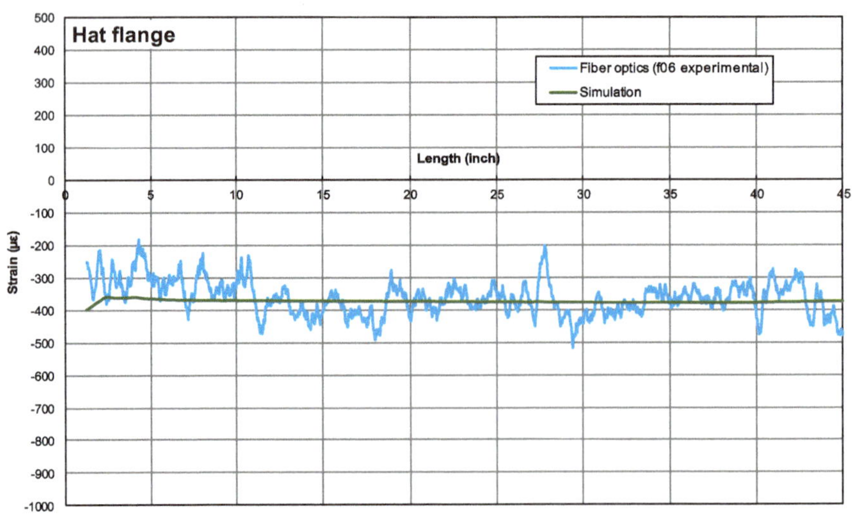

FIGURE 7.39 **Hat flange residual strain comparison.**

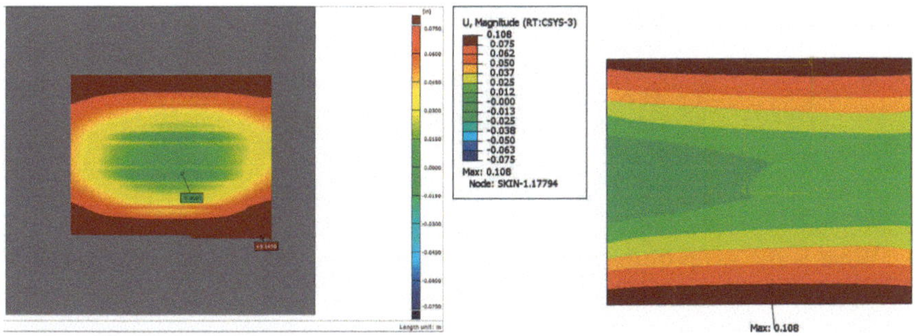

FIGURE 7.40 Hat panel warpage comparison.

simulation cross almost in the middle of the data collected from the experiments showing very close correlation. The values are much lower than those seen in the skin, which can be attributed to the geometry change across the length where the measurements are taken, but also the smaller thickness in that region. The data suggests that more attention should be provided to the skin/stiffener interface area compared to the actual hat section. Although it is worth noting that the running load does change depending on the shape and the ramps in any specific geometry, which should be considered during the design phase.

Finally, a comparison of the warpage measurements from the FEM model and experimental data was done. Note that the simulation utilized similar boundary conditions to ensure that the behavior is captured as accurately as possible. As shown in Fig. 7.40, the displacement fields match closely comparing both experiment and simulation. The simulation shows a wider distribution of the displacement near the edges. The max displacement values differ by approximately 28% but it is important to note that the overall behavior is much more important than the actual individual values. This data will be used to provide design and stress engineers' guidance to try and minimize this behavior and can also be used for tool compensation as required.

The analysis and experiments also provided insight into the flexibility of co-cure parts compared to a co-bonded or secondary bonded parts. The warpage values in this case seem to be small enough where applying force a finite amount of force to the panel we were able to bring it back to nominal shape without adversely impacting the structural part integrity unlike bonded parts which were shown to require more force to perform the exact objective.

With this information in mind the reader should have some clear understand of some of the issues that will be challenging during any project that deals with such parts and tools. The data presented here can aid in making the right decisions based on the application and situation when time arrives.

Chapter questions

1. Name four available commercial software's that can be leveraged for process modeling of composites and the purpose of each?
2. Define what CHILE is?
3. What are the steps that are taken to evaluate a composite part for producibility?
4. What is the main difference between using a geometry-based evaluation of a part and a physics based?
5. What are fiber optic sensors?
6. Name two of the main production concerns when it comes to tooling for composite structures?

References

[1] https://www.convergent.ca/products/compro-simulation-software.

[2] Johnston A, An intergrated model of the development of process-induced deformation in autoclave processing of composite structures PhD Thesis, University of British Columbia, 1997.

[3] https://www.compositesworld.com/news/darpa-presents-tff-program-for-rapid-low-cost-composites-to-replace-metals-in-small-parts-for-defense-applications.

[4] Nelson RH, Cairns DS. Prediction of dimensional changes in composite laminates during cure. 34th International SAMPE symposium and exhibition 1989;34:2397–410.

[5] Zobeiry N, et al. Efficient Modelling Techniques for Predicting Processing Residual, 2003. Available from: https://www.researchgate.net/publication/242157911_Efficient_Modelling_Techniques_for_Predicting_Processing_Residual_Stress_and_Deformation_in_Composite_Parts.

[6] Padovec Z. Springback analysis of thermoplastic composite plates. Applied and Computational Mechanics 2012;6:25–34. (2012).

[7] Bogetti TA, Gillespie JW Jr. Process-induced stress and deformation in thick-section thermoset composite laminates. J Compos Mater 1992;26(5):626–60.

[8] Sadd M. Elasticity, Theory, Application and Numerics. Elsevier Butterworth-Heinemann.; 2005. Available from: https://www.amazon.com/Elasticity-Applications-Martin-Sadd-Ph-D/dp/0124081363.

[9] http://www.roymech.co.uk/Useful_Tables/Beams/Curved_beams.html.

[10] Hasan Z, et al., Design, Analysis and Fabrication of Thick Co-cured Wing Structures", Composites Part B, 2019, in review. Available from: https://www.sciencedirect.com/science/article/abs/pii/S1359836818336291.

[11] Convergent Manufacturing Technologies (2013). RAVEN simulation software (Version 3.4.7). Available from: http://www.convergent.ca/products/raven-simulation-software.

[12] Dong A, et al. Cure cycle optimization of rapidly cured out-of-autoclave composites. Materials 2018;11:421.

[13] Optical Distributed Sensor Interrogator Model ODiSI-B: User's Guide ODiSI-B Software 5.2.0, 2017, LUNA.

[14] https://www.gom.com/metrology-systems/atos.html.

[15] Guo Z. Residual stresses characterization of laminated composites curing by FEM and FBG sensor. Poly Poly Comp 2011;19(2–3.).

[16] Kim. Y, White S. Stress relaxation behavior of 3501-6 epoxy resin during cure. Poly Eng Sci 1996;36(23):2852–62.
[17] Kim. Y, White S. Viscoelastic analysis of processing-induced residual stresses in thick composite laminates. Mech Comp Mater Struc 1997;4:361–87.
[18] Kim. Y, White S. Process-induced residual stress analysis of as4/3501-6 composite material. Mech Comp Mater Struc 1998;5:153–86.
[19] Prasatya P, et al. A viscoelastic model for predicting isotropic residual stresses in thermosetting materials: effects of processing parameters. J Comp Mater 2001;35:826–48.
[20] Ruiz E, Trochu F. Thermomechanical properties during cure of glass-polyester rtm composites: elastic and viscoelastic modeling. J Comp Mater 2005;39:881–915.

Chapter 8

Future lookout

Chapter outline

This chapter will go over some of the upcoming advances in the tooling technology primarily for aerospace applications and will discuss the author's thoughts on what should be anticipated in the future. The main driver for innovations in the tooling industry is mainly cost reduction and improved quality. That comes in the form of evaluating tool materials that can be quicker to procure, withstanding higher number of cycles and require less maintenance. Almost all the companies in the industry have a need for such advancements given the extreme competition among the different players. Universities and research institutions are constantly working on these improvements and the main themes that everyone is focused on can be divided into several categories:

- **Material advancements**: this involves looking into new materials and associated processes to achieve the set of requirements whether it be higher temperatures, ease of machining, ease of repair, etc.
- **Design philosophy**: this has to do with changing the way tool design occurs and implement methods and techniques that can minimize the drawbacks of legacy approaches. This can be done by ensuring that the tool design is thought through early on during the design phase and not waiting for a full part to be complete before considering how to laminate a part or assemble it.

1 Additive manufacturing

Additive manufacturing (AM) is an emerging technology that manufactures three-dimensional (3D) objects directly from digital models through an additive process, typically by deposition of successive layers of polymers, ceramics, or metal materials [1]. The use of additive manufacturing offers a feasible alternative for producing composite tooling with significantly shorter lead times and/or lower production costs [2]. Given the fabrication approach of deposition of the

Tooling for Composite Aerospace Structures. http://dx.doi.org/10.1016/B978-0-12-819957-2.00008-0

FIGURE 8.1 Tooling used for composite fabrication built from additive manufacturing [4–7].

material many geometries that would have been thought impossible to manufacture using legacy techniques are now possible via AM [3]. Examples of 3D printed tooling are shown in Fig. 8.1.

Generally speaking, there are seven recognized categories when it comes to additive manufacturing [8] and many other techniques that exist out there stem from one of these methods. All methods of additive manufacturing fabricate 3D parts in accordance to a pre-programmed computer-aided design drawings. These methods are:

- Stereolithography apparatus (SLA)
- Material jetting (MJ)
- Fused-deposition modeling (FDM)
- Powder bed fusion (PBF) (encompasses several AM techniques such as: selective laser sintering (SLS), selective laser melting (SLM), selective heat sintering (SHS), electron beam melting (EBM), direct metal laser sintering (DMLS), and direct metal laser melting (DMLM))
- Laminated object manufacturing
- Direct energy deposition (DED)
- Binder jetting

The use of these methods in tooling applications vary by industry and material but when it comes to the aerospace industry laser sintering is probably the most popular 3D printing method for the creation of durable metal parts for aerospace engines and fused deposition modeling (FDM) is mostly used for processing 3D-printed tooling applications.

TABLE 8.1 Advantages of AM versus conventional manufacturing approaches.

AM advantages	Conventional manufacturing advantages
• Economical for low production volumes • Less material consumption • Minimal machining cost • Lower capital investment • Less complex logistics	• Can support large production volumes • Easily processes/machined materials • Centralized manufacturing • More durable

The use of AM for aerospace tooling can be broken down into three general areas:

- Rapid tooling needed for master molds, or part lamination tools needed for low production rates.
- Production layup tools used to produce composite parts, either for autoclave or out-of-autoclave processes.
- Washout mandrels for fly-away tooling.
- Supplementary tooling including holding fixtures, jigs, trim tools, and other assembly tooling restricted by its size and load application.

The use of AM for production tooling offers a feasible alternative for conventional tooling methods which are time consuming and costly in the development of new products [9]. Given the rapid turnaround time compared to conventional methods it can deal with low-volume production and late-part design changes. Unlike conventional tooling that rely on subtractive fabrication approaches such as machining, AM creates the final shape by adding materials generating less waste and requiring lower production steps ultimately saving cost. A comparison of conventional versus AM is shown in Table 8.1.

Most recently a demonstrator thermoplastic composite mold was built at Oak Ridge National Labs Big Area Additive Manufacturing (BAAM) technology using a carbon-fiber filled Acrylonitrile-Butadiene-Styrene (ABS) thermoplastic [10]. The tool showed high durability following the demolding of 10 parts quantified by scanning the tool using a Laser 3D scanning with small variations. This offers another step towards getting AM mature enough to establish its place as a standard approach for building tools in the near future (Fig. 8.2).

With many of the advantages AM tools offer comes limitations to their use:

- The size of the tool is one inhibitor. The tool can only be of a certain size and shape at this point (of course depending on the fabrication approach) given the limitations of the machines that exist to support larger scale parts. But it should be noted that envelope restrictions do not limit tool size because a large tool or jig can be printed in segments and joined using thermal welding

(A) (B)

FIGURE 8.2 Additive manufactured mold; (A) As printed mold showing rough surface, and (B) The mold after surface finish and coating ready for producing composite parts [10].

or structural adhesive bonding methods (although not preferred given that anytime a joint exist there is potential for issues down the road).

- The use of AM is application dependent since they can only withstand specific temperatures. Low-temperature, short-run tools for prototyping that won't see high temperatures are commonly 3D-printed and well within AM's scope. But tools that must withstand higher temperatures and pressures of an autoclave might be more difficult to achieve using AM tools.
- The negative impact of CTE variation between tool and part. With the availability of different materials for the different fabrication approaches, the designer needs to be aware of the materials selection for the application. Ultem 1010 reportedly has the lowest CTE, and a T_g of 217 °C which is a great candidate for use with composite molds. Another way for dealing with CTE is to avoid female tool designs, if possible, which can lock the part in the mold, and instead opt for mandrels, or male tools.
- The surface finish and tool porosity. With sanding and surface finish, the surface can be machined to a specific surface finish (e.g., 32 RA) and in order to get it to a smooth surface finish a sealer can also be used for that purpose.

2 Tool-less assembly

For most of the assembly approaches that exist to date we use a significant number of tools to enable the final assembly of whatever product we need. When speaking about composites, another level is added since we use more bonding at different stages compared to metals adding to the number of tools needed. Composite parts require several tools/fixtures in order to complete a finished product that are summarized below:

- Lamination molds
- Trim fixtures
- Bond fixtures
- Assembly fixtures
- Work holding fixtures

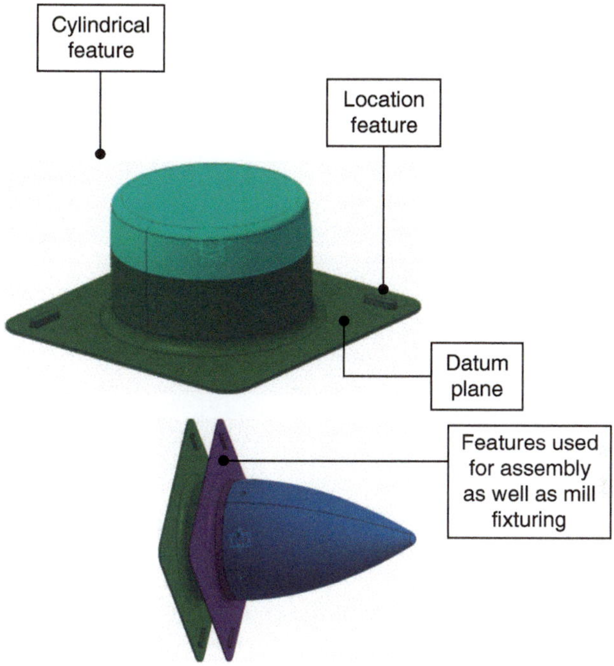

FIGURE 8.3 Examples of features used in tooling to simplify the assembly process.

In order to reduce labor hours, tooling number, and tooling cost, one remedy to this would be to include smart features in lamination molds in order to streamline the manufacturing process. Some of those features include:

- Pin holes
- Locating features
- Trim features
- Flat planes or cylindrical features

These common features will enable common work holding, locating, and assembly locations throughout the manufacturing process. Examples are shown in Fig. 8.3. Advantages they offer include:

- Significant labor reduction
- Reduction of processing steps
- Higher percentage of quality approval

3 Self-heating tools

One of the biggest bottlenecks when it comes to the production of composite parts is the availability of autoclaves and/or ovens needed for cure. An average thermoset composite part requires around 8 hrs for a part to reach full degree

FIGURE 8.4 Large scale autoclaves used to fabricate composite parts [11].

of cure and depending on the size of the equipment used, you can only fill so many parts for restriction of cure cycle differences or size. The cost of these autoclaves can reach millions of dollars, which limit the supply chain of where certain parts can be fabricated. As an example of a very large autoclave is the one that will be used to build the 777X wing which is one of the largest single piece composite parts ever to be built in the aerospace world. A figure of the autoclave during transportation is shown in Fig. 8.4.

One solution to that problem is offering tools that are self-heated as shown in Fig. 8.5. These are tools that can be used to laminate a composite part and at the same time use to heat outside an autoclave where heat is provided via several techniques one of which is direct injection of hot air inside tubes to be distributed around the part for temperature. If pressure was needed, the tool can

FIGURE 8.5 Self-heating tools used for composite fabrication [12].

be achieved by locating it under a press or if no pressure is required (e.g., OOA material) the heat will be sufficient.

There are many limitations to this technique including size and part thickness. As the thickness increases, it is more difficult to get uniformly distributed heat across the part and the ability to achieve the desired final cure might be questionable. Another issue is reliability. Since the heating relies on added complexities associated with the heating elements, we might anticipate issues with the system compared to the autoclave. This approach has yet to become mainstream especially for building aerospace parts but has a lot of potential for specific parts and should be considered as part of future advancements.

4 Tool-less part fabrication

Given the ever-increasing use of thermoplastic materials prompted by the advantages they offer as discussed in Chapter 3 and 7, engineers have also been looking at ways to take advantage of those properties to aid in simpler tooling approaches as well.

One unique proposal to fabricate parts from thermoplastics is to avoid the need for tooling all together. The tool-less process uses two 6-axis robots working cooperatively to place thermoplastic tape into open space within a metallic or similar frame that provides the boundaries of the structure being fabricated. One robot consists of a standard unidirectional tape placement system that provides laser heating to perform in-situ consolidation of the thermoplastic material. The second support robot works directly *opposite* the automated tape layer and consists of a flat metallic surface, providing, in effect, a *movable* tooling surface against which the ATL places its tape. The tape head and the support head thus move together through 3D space, placing material. Each end of each tape placed is anchored to the frame, which can assume a variety of shapes, depending on the application. Further more, the tape can be manipulated by the robotics to change direction within the 3D space to build contoured and complex shapes. An overview of this process is shown in Fig. 8.6. For this technology to

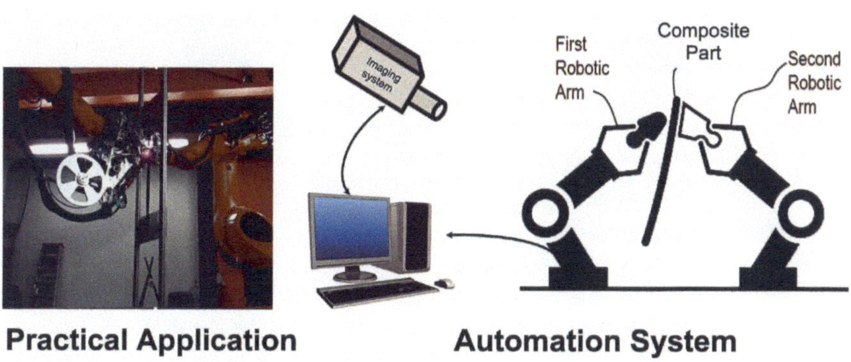

FIGURE 8.6 Tool-less fabrication approach schematic and demo in progress [13].

come to life many hurdles need to be overcome but its one step toward an industry with no tools needed for composite production, which is exciting.

5 Future predictions

The pursuit for improved materials for aerospace parts and tools is never ending. The primary drivers are lower weight, improved performance, temperature and fatigue resistance, and lower acquisition, and operation costs. One of the most significant current barriers to using new materials is the qualification aspect.

Given the amount of uncertainty related to the behavior of composites under different conditions, engineers have developed a process to evaluate the material and designs in a systematic way, which is known as the building block approach [14] which was discussed in several chapters early on in this book. The process follows a step-wise method, initially the material properties and allowables are computed which involves testing thousands of coupons. This is followed by element testing for specific details used in the structure (e.g., stringer terminations) that consider smaller scale components under load. The numbers of tests at this stage vary by design complexity and the amount of perceived high-risk details. Sub-component and component level testing is considered next. These tests account for larger-scale details and represent the actual structure in the aircraft. Finally, a full-scale article is tested. This approach reduces the risk by understanding the facts at the earlier stages of the design rather than waiting until the final design and discovering an issue which can be very costly to fix and can jeopardize the entire program. Fig. 8.7 shows a schematic representation of both a structural and manufacturing building block diagram.

In order to overcome those drawbacks in the building block approach many researchers have recommended different solutions [14] most notably is the use of software-based analysis methods and, specifically, the introduction of

FIGURE 8.7 Current conventional building block schematic for both the structural testing and manufacturing approach.

multi-scale analysis to develop allowables and reduce deign risks. These analysis tools offer engineers the ability to offset physical testing, reduces the time and cost of research and development which in turn expedites the material qualification and design solution verification.

Some of the benefits of this method are:

1. Reduction in the number of structural tests required which will save time and cost. Instead we can rely on simulations to provide the data that will supplement the testing.
2. Evaluate specific details that impact the structural performance that are difficult to be executed via testing.
3. Reducing the conservatism in the analysis methods. This can translate into lower structural weights.
4. Accelerate the insertions of materials and novel design solutions into the aerospace industry.

These types of multi-scale analysis have been mainly used for structural analysis of parts and the focus on tooling and assembly has not seen as much attention. One of my interests and hopes in the next several years is to advance such tools for the purpose of understanding the entire fabrication approach for composite materials starting from the material itself, to the lamination, to the cure, to the assembly and operation. Such analysis tools have the potential to prevent issues during the early phases of any program and reduce the cost significantly by minimizing the testing to a finite subset rather than relying on testing for all variables as is done nowadays for all aerospace products.

Fig. 8.8 shows an outline of the process flow for future advancements in analysis of tooling and assembly.

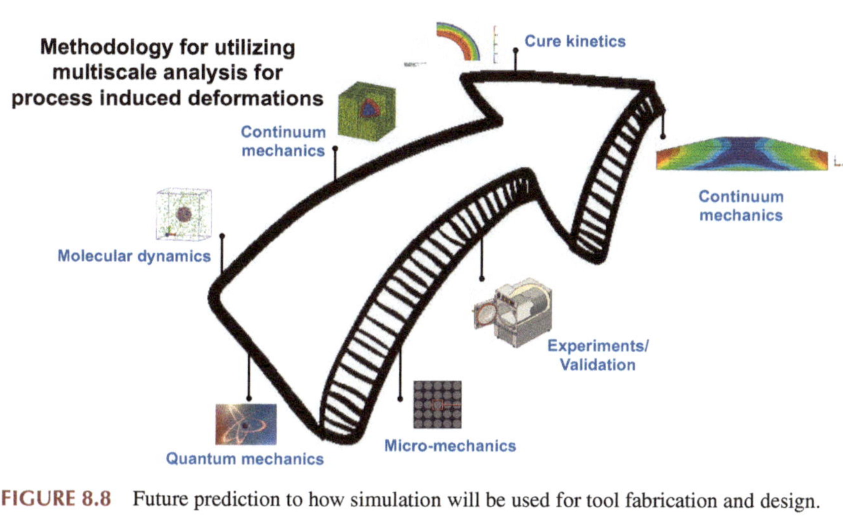

FIGURE 8.8 Future prediction to how simulation will be used for tool fabrication and design.

Chapter questions

1. Name three advancements happening now in the aerospace tooling and part fabrication industry?
2. What is meant by "multiscale" analysis?
3. What are some of the disadvantages in self-heating tools?
4. Name five benefits of using AM tools?

References

[1] Ford SLN, Additive manufacturing technology: potential for US manufacturinng competitiveness, J Int Comm Econ **6**, 2014, 1–35.

[2] Thierry R, Striukova L. From rapid prototyping to home fabrication: how 3d printing is changing business model innovation. Technol Forecast Social Change 2016;102:214–24. *Crossref*, doi:10.1016/j.techfore.2015.07.023.

[3] Bourell D, et al. Materials for additive manufacturing. CIRP Annals 2017;66(2):659–81. ScienceDirect, doi:10.1016/j.cirp.2017.05.009.

[4] http://www.owi-lab.be/content/3d-printing-large-components-offshore-renewable-energy-industry-inspiration-future-cost.

[5] https://www.stratasys.com/tooling/composite-tooling.

[6] https://www.3dprintingmedia.network/smartech-top-composite-additive/.

[7] https://www.proboat.com/2018/09/3d-printing-of-a-construction-mold/.

[8] "The 7 Categories of Additive Manufacturing." The 7 Categories of Additive Manufacturing | Additive Manufacturing Research Group | Loughborough University, Loughborough University, https://www.lboro.ac.uk/research/amrg/about/whatisam/.

[9] Rajaguru J, Duke M, Au C. Development of rapid tooling by rapid prototyping technology and electroless nickel plating for low-volume production of plastic parts. Int J Adv Manufact Tech 2015;78(1–4):31–40.

[10] Hassen, et al, The durability of large-scale additive manufacturing composite molds. https://www.researchgate.net/publication/324391116_The_Durability_of_Large-Scale_Additive_Manufacturing_Composite_Molds.

[11] https://www.boeing.com/company/about-bca/washington/first-777x-autoclave-busts-a-move-to-the-future-at-everett-site-09-08-2015.page.

[12] https://www.slideshare.net/Gregleidig/out-of-autoclave-tooling.

[13] https://www.compositesworld.com/news/general-atomics-aeronautical-developing-tool-less-thermoplastics-composites-process.

[14] Hasan Z, The impact of multiscale analysis on the development of virtual material allowables. In: Conference: The Composites and Advanced Materials Expo; 2018. https://www.researchgate.net/publication/331333024_The_Impact_of_Multiscale_Analysis_on_the_Development_of_Virtual_Material_Allowables.

Appendix

Spring-in simulation example

The following example introduces the reader to the process of simulating a cure cycle for a composite part similar to that shown in Fig. A1. This type of geometry is commonly referred to as spar and is used in wing structures, fuselage sections, and other areas across an airframe. In order to perform the simulation the reader needs to have the finite element software Abaqus and the plug-in Compro offered by Convergent Manufacturing Technologies Inc. installed on their computer or laptop. For additional details on each software the reader is referred to the website of each corresponding developer that can be found online.

The part geometry used for this example is a section of a full-scale part. The composite material used is 0.322-inch thick Cycom 5320-1 resin with IM7 12K fibers. The tool is simulated as a 0.35-inch thick aluminum tool. Depending on the size and shape of the geometry the analyst is recommended to evaluate the best way to model it and the type of assumptions they shall consider. The purpose of this example is to provide students, educators, and professionals a step-by-step example on how to model this type of structure and analyze it for part thermal gradients under a specified cure cycle and spring-back due to accumulated residual stresses.

The results of this analysis can be used to perform changes to a design as needed based on a trade study of different variables or substantiate a decision made for production support purposes. This skill can be very valuable for students to support research and development activities for future work.

This example will first discuss the model setup in Abaqus and will primarily focus on two analysis steps:

- Thermo-chemical required to evaluate the thermal distribution in a part, and
- Stress-deformation that is used to predict the deformation and spring-back values.

The part considered in this example is shown in Fig. A2. Note that the example discussed here can be applied to any design with appropriate changes as the analyst deems necessary. All the information presented here is for educational

Tooling for Composite Aerospace Structures. http://dx.doi.org/10.1016/B978-0-12-819957-2.00014-6

FIGURE A1 Example of a composite spar part used on aircraft structures.

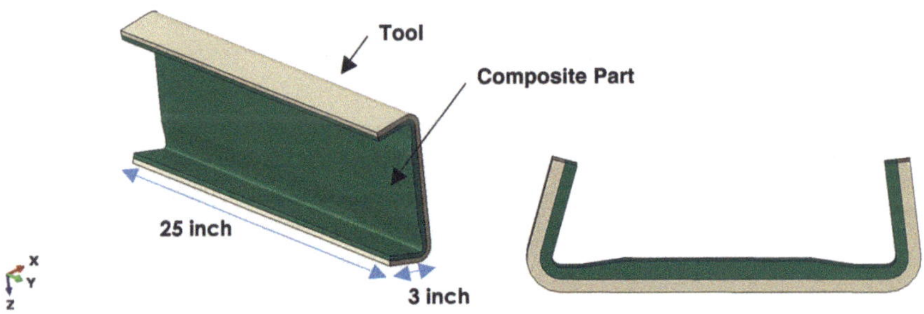

FIGURE A2 Geometry used in the following simulation example.

purposes only and the reader must validate all materials data, properties, and validation of the models prior to using it for any specific design.

1 Model set-up

Step 1. Every simulation always starts with a model that is built using any computer aided design (CAD) software. Fig. A3 shows a snap shot of the part model that was made using Solid works in this case. Other software's that can be used are CREO, NX, CATIA, which are typically used in the aerospace industry. Solidworks on the other hand is typically used in academic settings due to its inability to deal with large assemblies as effectively. In general, any available software will work if the part design can be made with the appropriate details include in the model.

Step 2: Once the CAD part is available import it into Abaqus software as shown in Fig. A4.

Under file select Import → Part

Note that you can also model simple parts in Abaqus but it is recommended to use CAD software in order to ensure the actual design details are reflected in the part. The imported part is seen in Fig. A5.

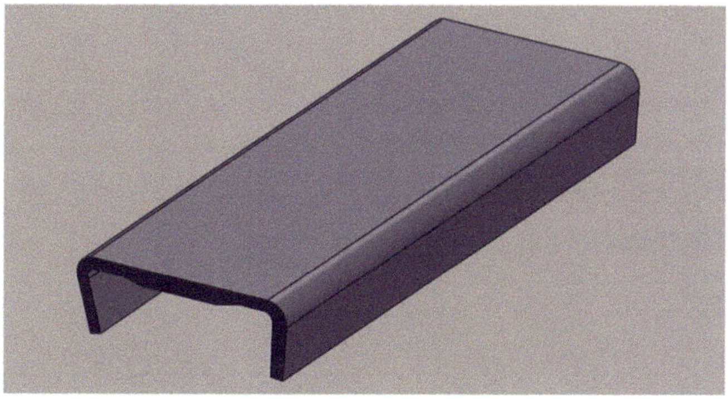

FIGURE A3 CAD model using SolidWorks.

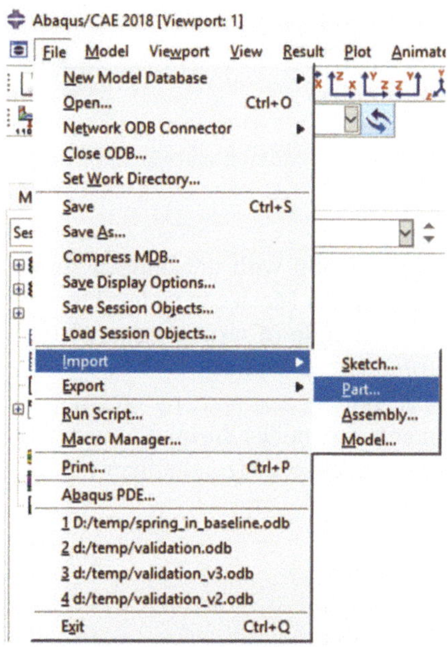

FIGURE A4 Import of part from CAD into Abaqus.

FIGURE A5 Imported part into Abaqus from SolidWorks.

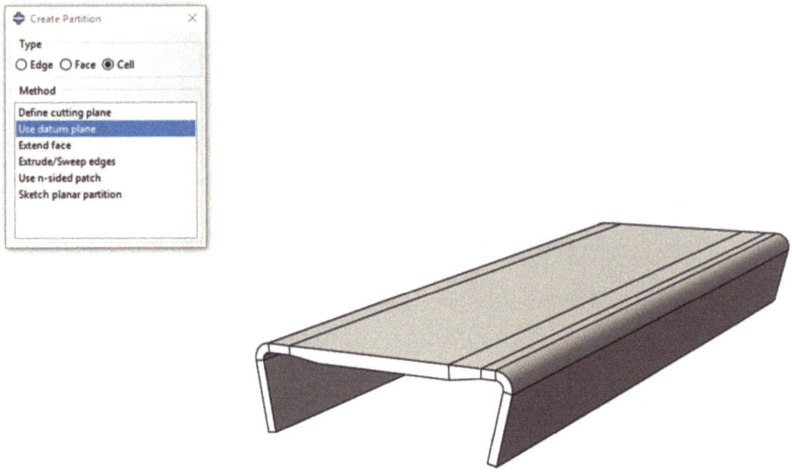

FIGURE A6 Part with partitions along critical features in the part to aid in the meshing process.

Step 3: Partition the part to help with meshing as shown in Fig. A6. Several other software's including Hyper Mesh can be used for meshing the part as well which typically has a better suite of tools compared to Abaqus especially for parts that have more complicated details.
Step 4: Go to the mesh module and start by creating seeds along the part in order to align the elements and nodes during the meshing process as shown in Fig. A7. The smaller the global seed size, the finer the mesh will turn out to be.

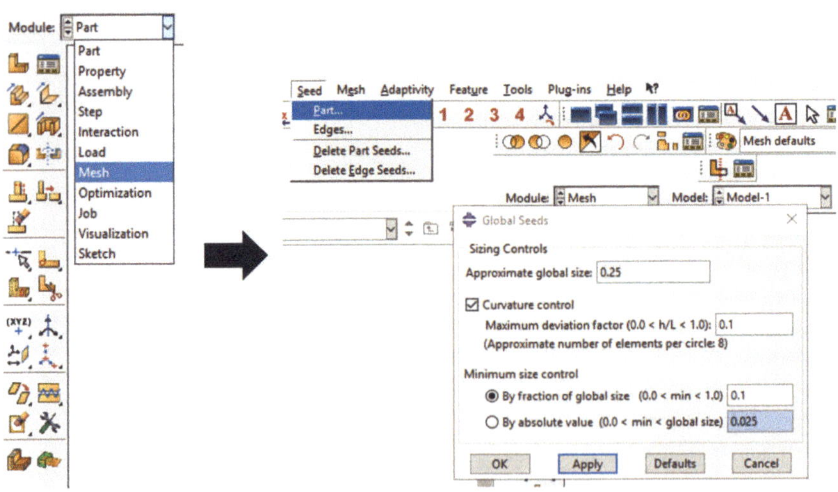

FIGURE A7 Applying global seed to the model prior to starting the meshing process.

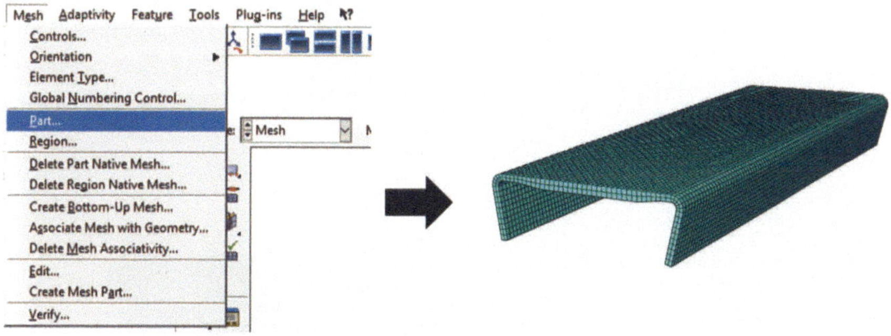

FIGURE A8 Part meshing process.

Step 5: After assigning the seed to the part select the "Mesh" tab and select mesh part as shown in Fig. A8.

Step 6: Transfer the part into an orphan mesh as shown in Fig. A9. An orphan meshed part provides the ability to have elements and nodes without the associated geometry. This allows for ease of manipulation of the part as needed. This step is typically not needed when you perform other types of FE simulations but since we want to also simulate a male tool it is easier to simply offset elements from the top part surface to create the tool than create a whole new part in CAD and mesh it independently. More detail on this will be shown in the next several steps.

Step 7: This step involves the generation of the tool part by performing edits to the original orphan mesh part. Go to "Edit Mesh" and offset the elements on the surface of the part as shown in Fig. A10. Select 0.35 inch as the total thickness and 2 elements through the thickness.

Step 8: Create a copy of the part that was just created and rename it to "Tool" or any other name you choose as shown in Fig. A11.

Step 9: Delete the first 2 row of elements from the bottom representing the part (ensure that the "delete associated unreferenced nodes" is checked as shown in Fig. A12). This will only keep the elements representing the tool. Repeat the same steps in order to remove the tool elements from the top on the part that was copied over that will represent the composite part.

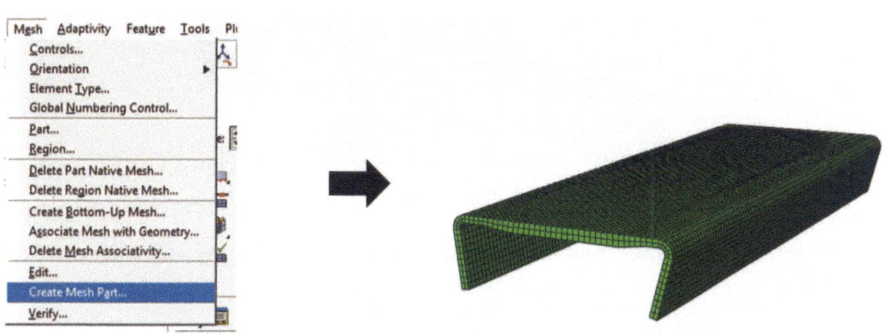

FIGURE A9 Create an orphan mesh by selecting the "Create Mesh Part" from the dropdown.

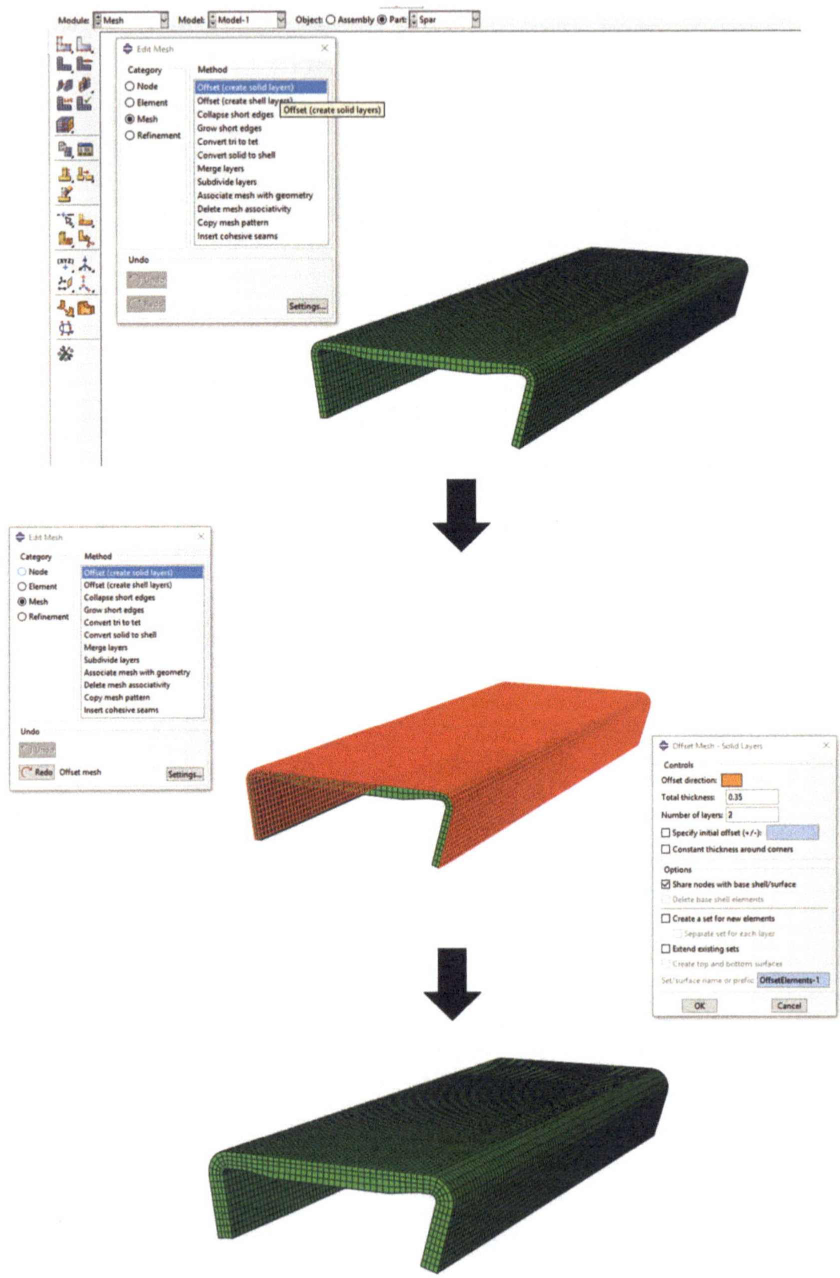

FIGURE A10 Sequence of steps to create an offset of the elements from the original part that will act as the tool part.

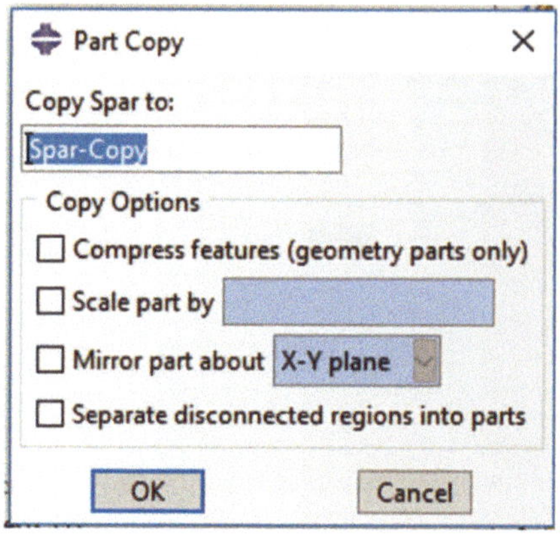

FIGURE A11 Copy the part that has the offset elements. This copy and the original part will act as the tool and composite part, respectively.

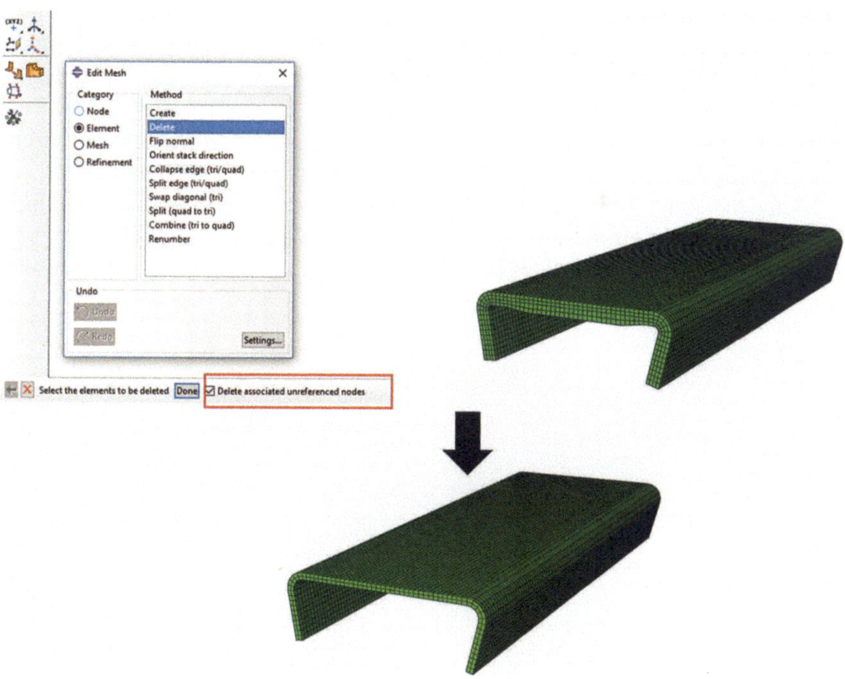

FIGURE A12 Steps to delete elements/nodes associated with the composite part and keeping the elements/nodes representing the tool. Same steps shall be done to keep the elements/nodes representing the composite part to be done on the part that was copied over.

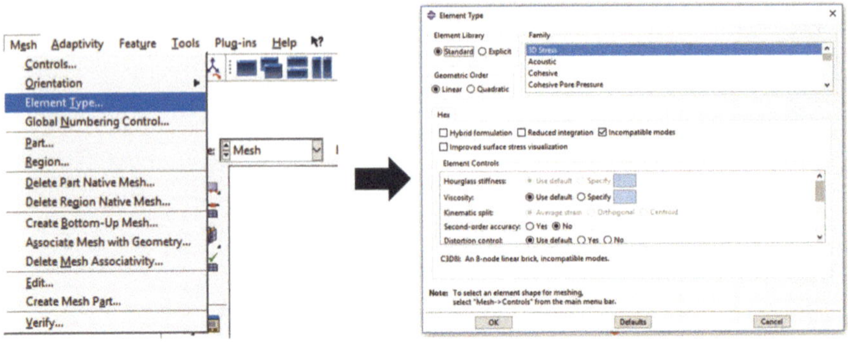

FIGURE A13 Assign element type to the part and tool.

Step 10: Assign the appropriate element type as shown in Fig. A13. In this case it needs to be an 8-node linear brick incompatible mode C3D8I element. This is the element type that is supported by the plug-in Compro which is needed for this type of simulation as you will see shortly.
Step 11: Assign material orientation. We select a "discrete" definition. The "Surface" selected should be the exterior surface to the part or tool and the 3-direction should be pointing outward as shown in Fig. A14. For more information on these type of simulation definitions for composite materials in Abaqus the reader is referred to the Abaqus user manual. This material orientation assignment is not needed for the tool part as the tool is assumed to be isotropic.

After meshing, assigning element type, and assigning orientation the next steps involve the definition of the different materials needed for the simulation and the associated sections. This is where the use of the Compro plug-in comes in.

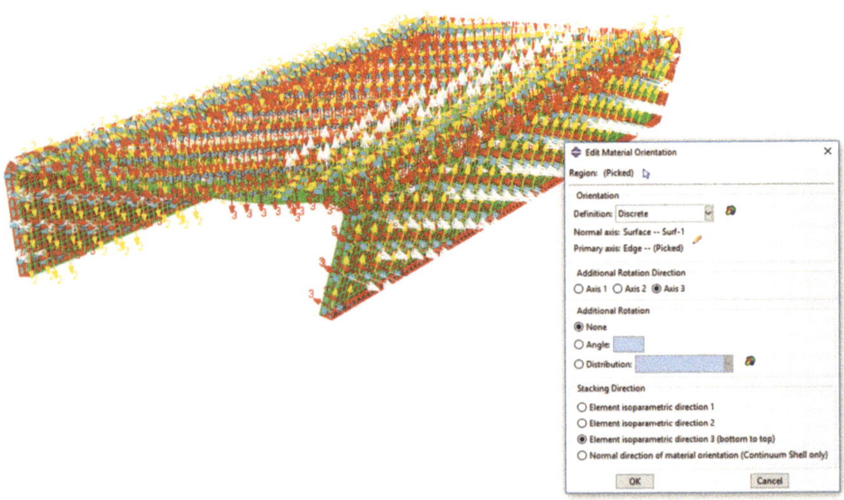

FIGURE A14 Material orientation assignment for the composite part.

Compro offers a suite of materials that have been characterized and fitted to ensure compatibility with the CHILE constitutive relation that was discussed in Section 7.2.2 Warpage Analysis back in Chapter 7. These properties span the different temperatures that we expect the part and tool to operate in and provides a more accurate response than using temperature independent material properties and the built in constitutive relations (e.g., Hook's Law).

Step 12: Open the Compro plug-in as shown in Fig. A15 and select the materials that are needed for the model setup. Make sure you select the correct model if you have several models in the analysis tree.

Step 13: Once a material is defined we need to create a section in order to assign to the different parts. For the purpose of this example we will need one section for the composite part and another for the tool as shown in Fig. A16.

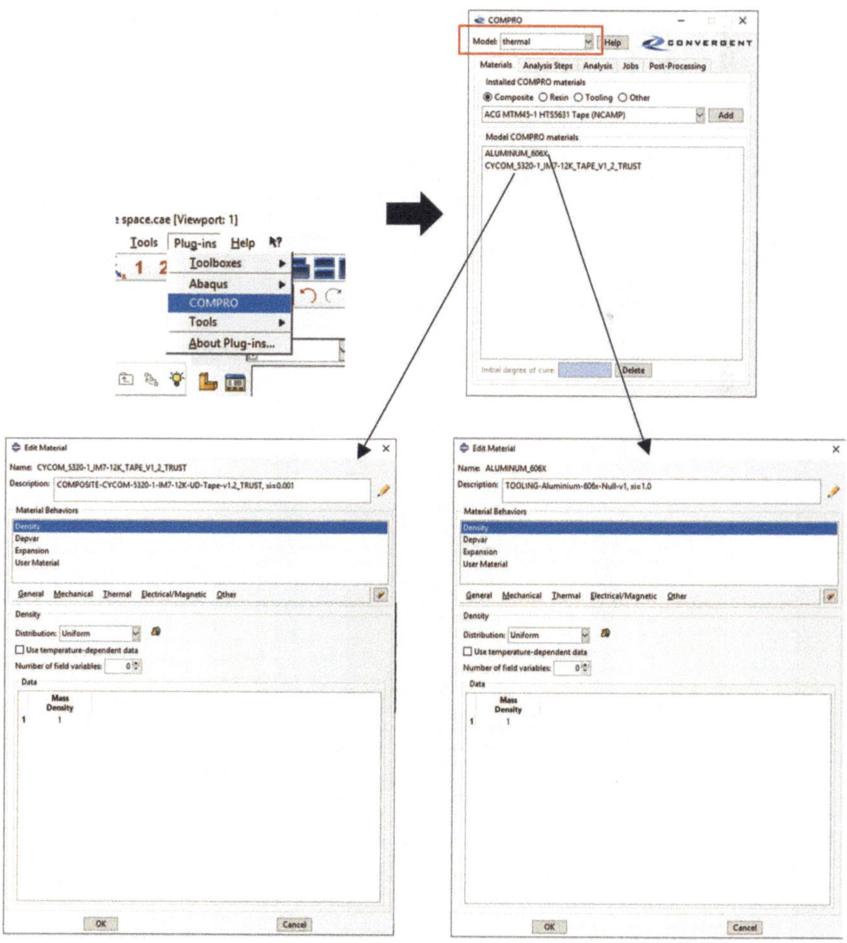

FIGURE A15 Material definition using the Compro plug-in.

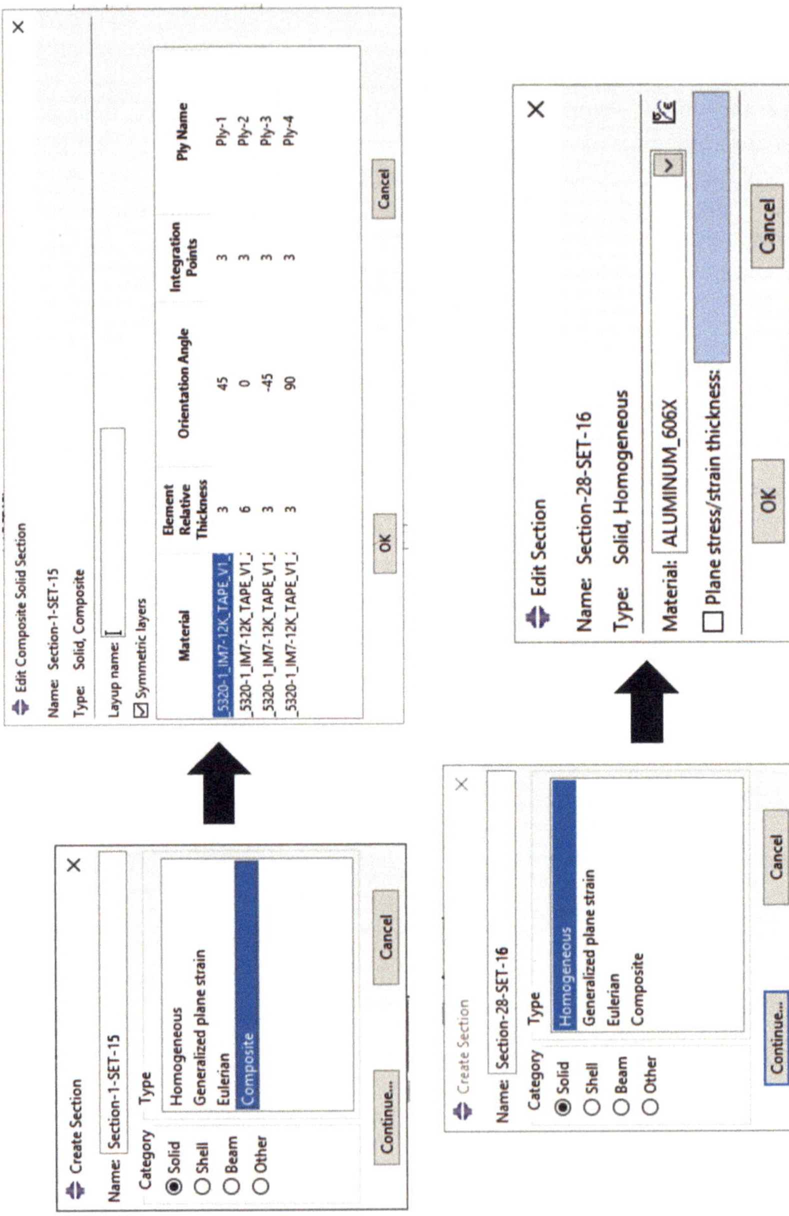

FIGURE A16 Section definition for the composite part and the tool.

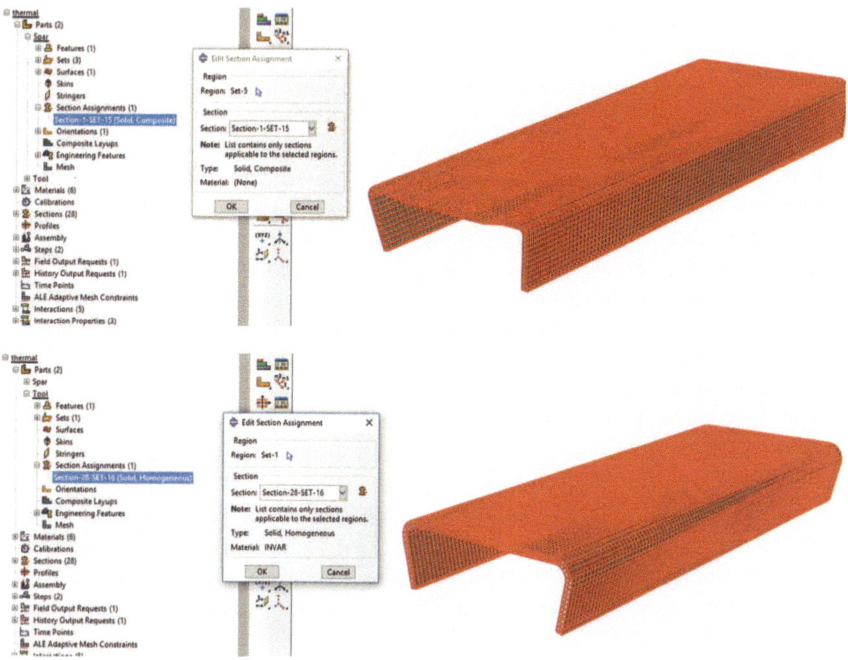

FIGURE A17 Section assignment for the composite part and the tool.

Step 14: Assign the different sections to the parts as shown in Fig. A17.
This concludes the part definition that is always needed as part of any FE simulation. We now focus on creating the assembly and applying all the necessary boundary conditions, loads, and running the jobs for analysis.
Step 15: Go to the assembly module and bring in both parts (part and tool) as shown in Fig. A18.

1.1 Thermo-chemical analysis

The thermo-chemical analysis will enable us to evaluate the temperature distribution across the model under a specified cure cycle instantaneously. In general practice, this is done by conducting a thermal profile test for each part in order to determine the thermal distribution via experiments and based on the results the lead and lag thermocouples are identified and used to track the cure during production. The distribution will depend on the material, geometry, tool, cure cycle, autoclave, and so on. By simulating this virtually it allows us to understand the behavior without conducting any expensive testing and adjusting as needed early on during the design.

This simulation needs to be initiated using the Compro plug-in since it will translate the element type automatically once the job is run from a stress deformation type to a thermal deformation, which is needed for this type of

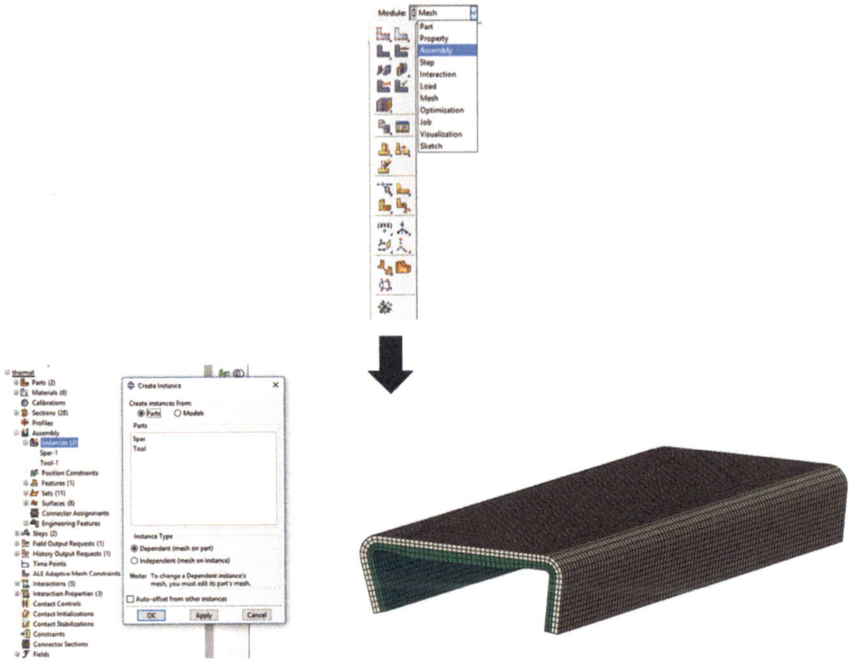

FIGURE A18 Assembly module.

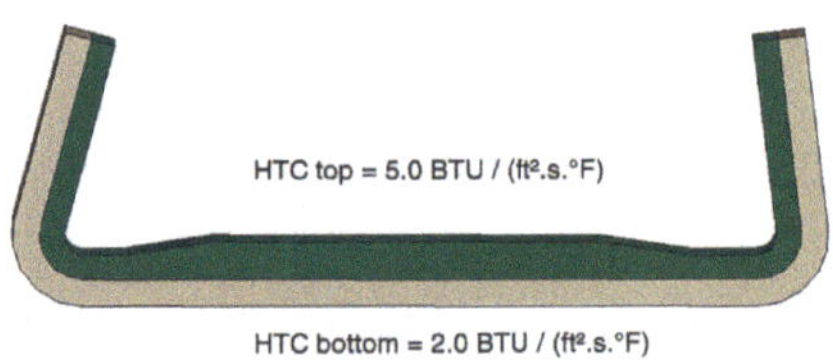

FIGURE A19 HTC values assumed for the model.

simulation. Convective heat transfer boundary conditions are applied to the sides of the part (top and bottom). Any thermal lag that we expect to see due to the thermal lag of the composite material and the tool material relative to the applied temperature cycle due to heat transfer conditions shall be captured in the results. In addition, due to internal heat generation during cure, any temperature overshoot will be captured using this analysis as well. The results of the analysis will act as input to subsequent stress-deformation as needed. Fig. A19 shows the assumed heat transfer coefficient (HTC) used for the different surfaces in the model. Some recommended values that can be used for simulation depending on the heat flow in the autoclave or oven are shown in Table A1. These values are for reference only and the reader shall adjust based on the details of their actual application.

TABLE A1 Reference HTC values that can be used based on temperature flow conditions.

Flow condition	HTC (BTU / ft^2.s.F)	HTC (BTU/in^2.s.F)
Stagnant Gas	1.7	0.0118
Low	2	0.0139
Medium	5	0.0347
High	10.5	0.0729
Impinging	18	0.1250

Step 16: Create a thermal-chemical step. This is done by opening the plug-in and selecting "Thermo-chemical" step as shown in Fig. A20. The time period represents the number of seconds we anticipate for the cure cycle. In our case we will use 31022 s (8.617 h).
Step 17: Create a node set called "ALL NODES" and pick all the nodes in the model as shown in Fig. A21. This node set will be used to assign an initial temperature of 70°F to all the nodes in the model as shown in Fig. A22.
Step 18: Create an amplitude table that includes the details of the cure cycle as shown in Fig. A23. In this example we start at 70°F and ramp to 250°F and hold for 120 min, this is followed by another ramp to 350°F and hold for another 120 min and then ramp down to 70°F again.

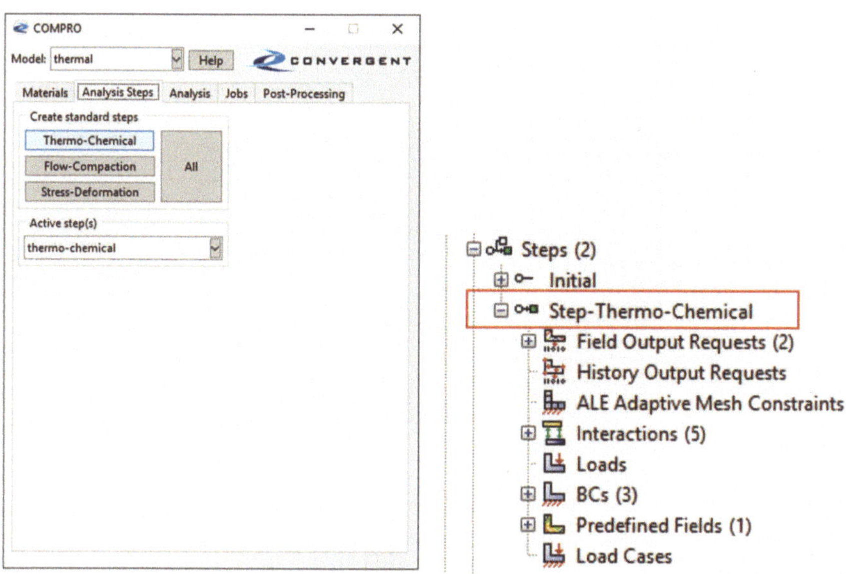

FIGURE A20 Creation of a thermo-chemical step.

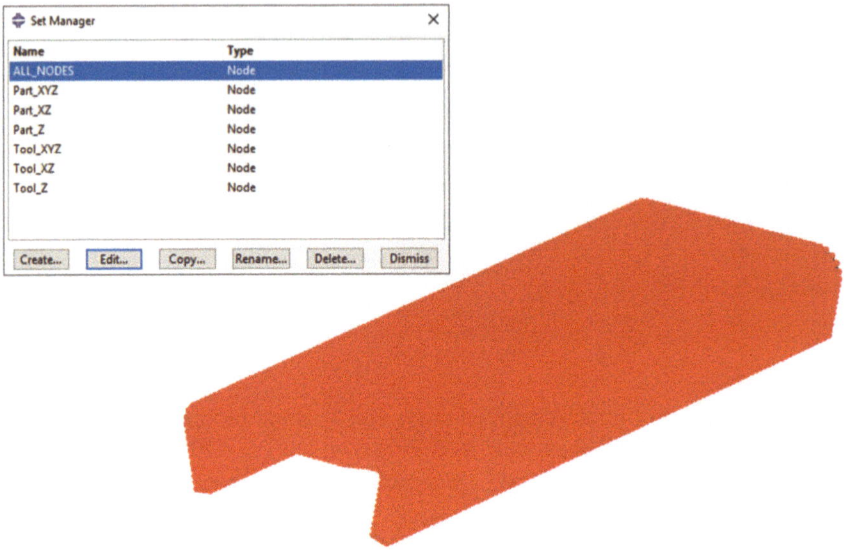

FIGURE A21 Assigning all the nodes in the model to a node set called "ALL_NODES."

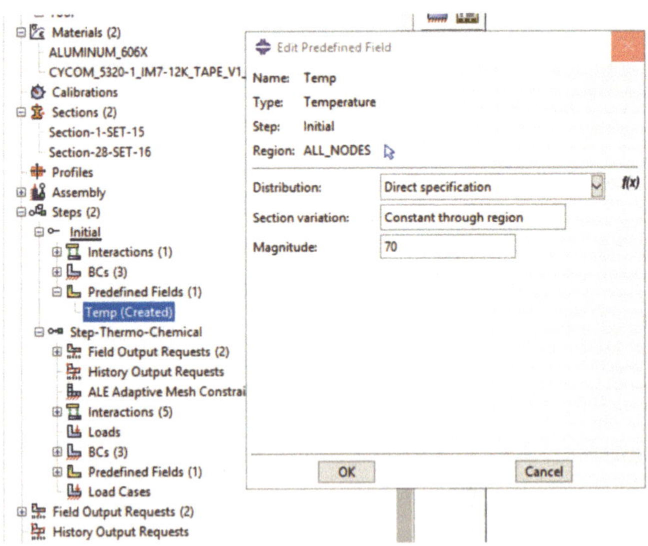

FIGURE A22 Under the initial step select "Predefined Fields" and assign a temperature 70°F to the "ALL_NODES" node set.

Step 19: Create several surfaces in order to assign the convective heat transfer boundary conditions to. In this case we will create two surfaces one for the top surface and another for the bottom surface as shown in Fig. A24.

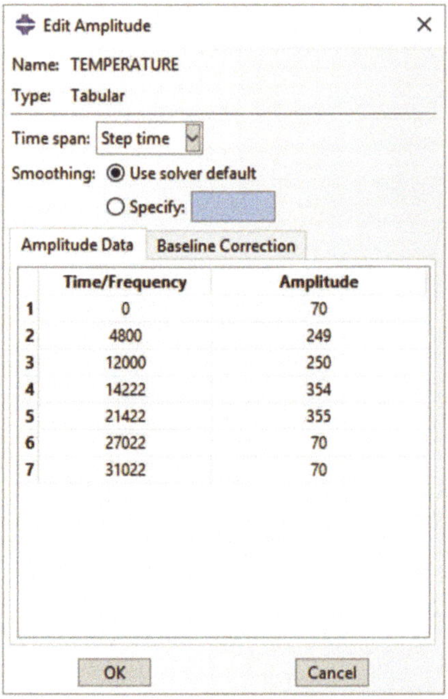

FIGURE A23 Temperature amplitude representing the cure cycle for this simulation.

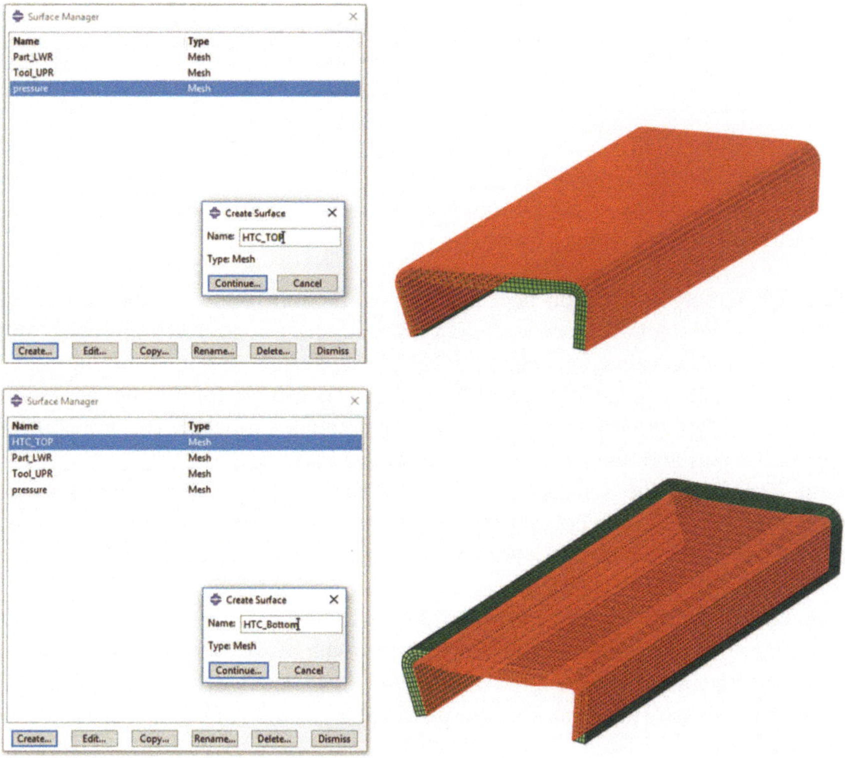

FIGURE A24 Creating both a top and bottom surface that will be used for the HTC boundary condition application.

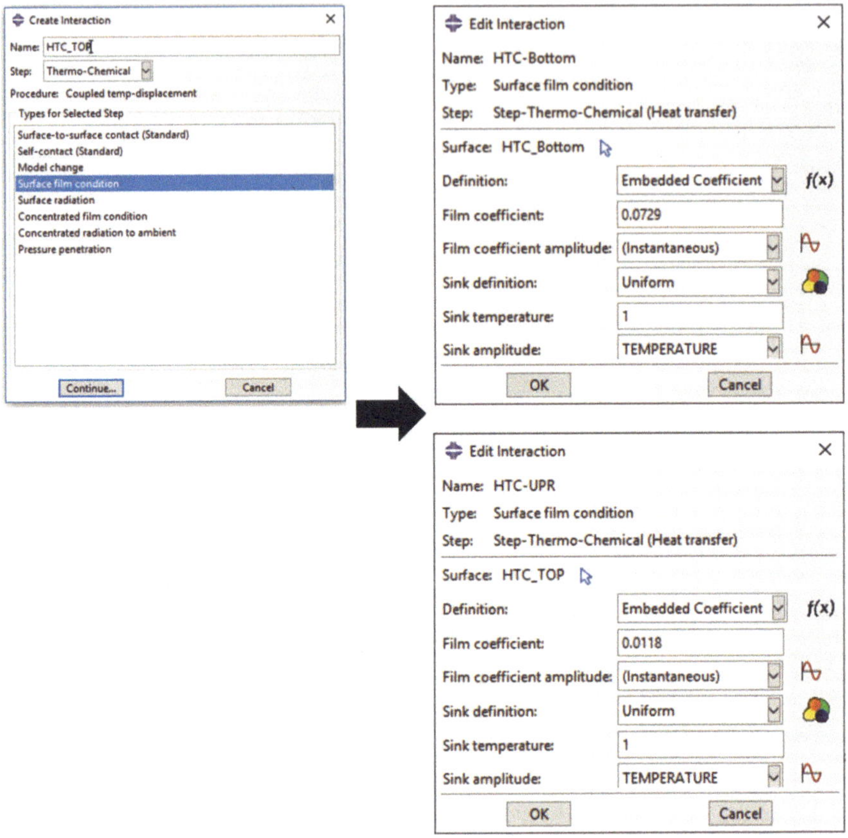

FIGURE A25 Assigning a convective heat transfer boundary condition on the part based on the expected heat flow in the autoclave during cure.

Step 20: Create convection heat transfer interaction in the Thermo-chemical step. Go to interaction module. Select "Surface film condition" and apply the appropriate HTC value and select the applicable surface as shown in Fig. A25.
Step 21: Create two surfaces one for the part and another for the tool as shown in Fig. A26. These surfaces will be used to create a surface-to-surface contact among the part and tool to simulate that they are attached.
Step 22: Create the surface-to-surface interaction between part and tool as shown in Fig. A27. The contact interaction property represents the behavior of the contact surface and the details of what we used for this example is shown in Fig. A27.
Step 23: Select the appropriate output fields for the model. One unique aspect here since we are using the plug-in Compro which uses many state variables representing temperature, degree of cure and other details we need to select the "State/Field/User/Time" in order for us to see the results. This is highlighted in Fig. A28.

Step 24: Run the model by opening the Compro plug-in and selecting the "Analysis" tab as shown in Fig. A29.

1.2 Thermo-chemical analysis results

After setting up the model and running the simulation it is very possible to get errors as is the case with every type of simulation and analysis. Ensure to debug the errors closely by referring to the Abaqus manual and the references provided by Compro. In this example, we are mainly interested to see the temperature distribution of the part during cure and if there are any major issues that need attention before continuing on with the stress-deformation analysis which will evaluate the spring-back in the model and resulting residual stresses. Fig. A30 shows the model at the end of the cure cycle and the temperature across the model is equal to 70°F as we would expect based on our inputs made earlier on.
Step 25: The plug-in provides the user the ability to plot the maximum and minimum temperature variation across the part during the cure as well as the

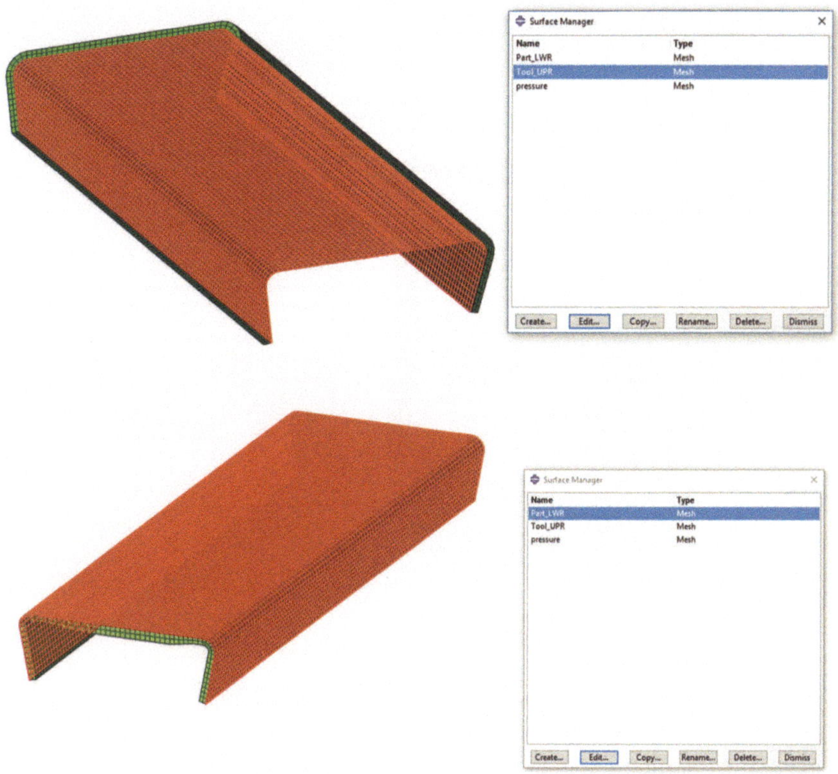

FIGURE A26 Since the tooling approach used here is a male tool we will select the top of the part to be in contact with the tool bottom surface.

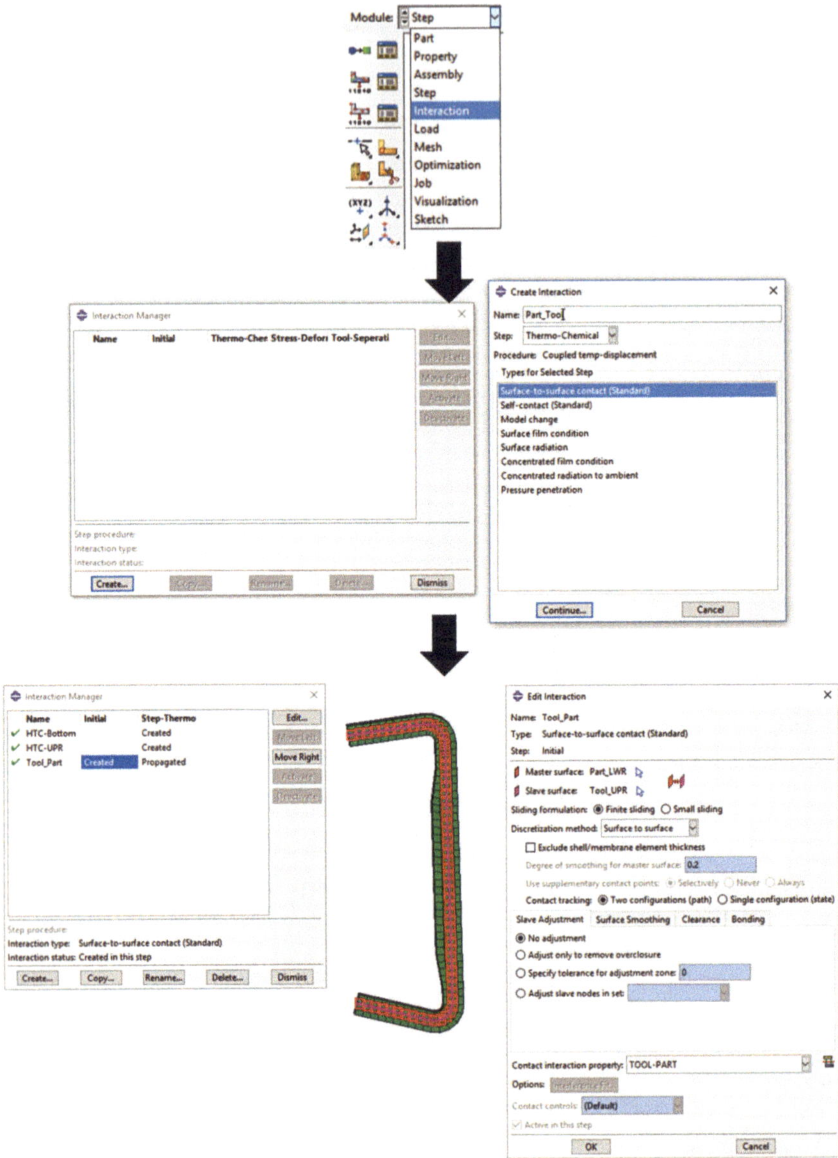

FIGURE A26 *(Continued)*

DOC. Fig. A31 shows the steps needed to generate the plots which are shown in Fig. A32.

As shown in the figures, we do not see any large variation or temperatures overshoots in the model which allows us to proceed with the next step of the analysis.

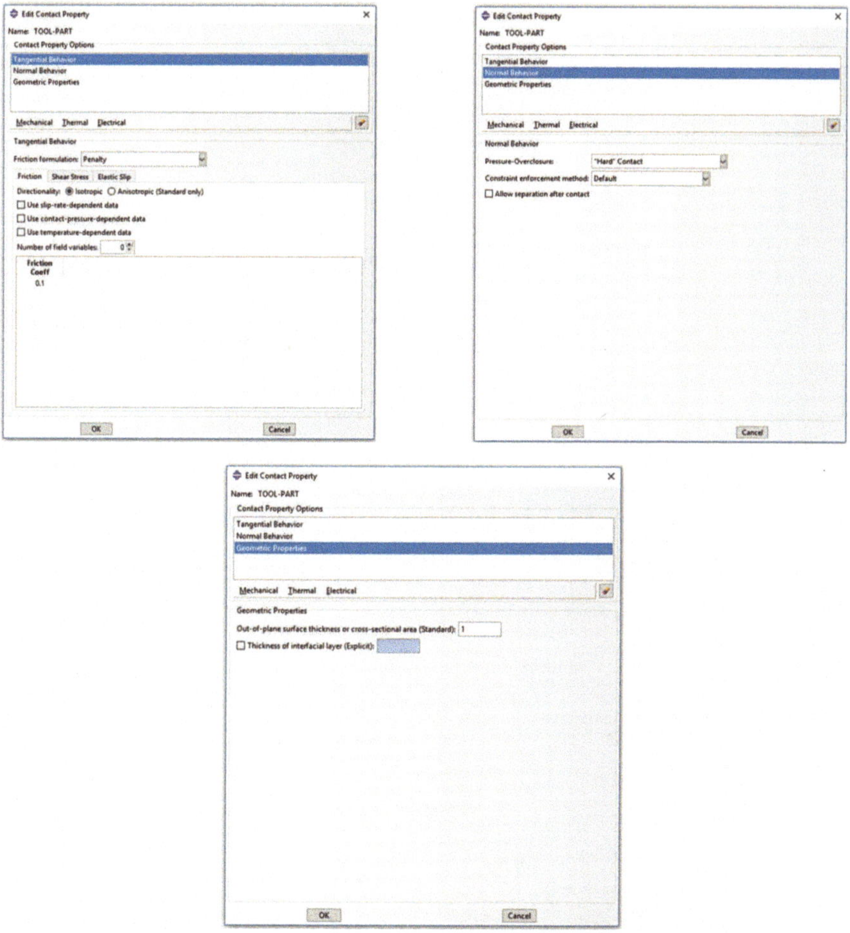

FIGURE A27 Contact interaction property details used in the simulation.

1.3 Stress-deformation analysis

For the stress-deformation analysis one of the first steps will be to apply displacement boundary conditions to the corners of the model. At the end of the analysis, the part is removed from the tool and allowed to deform freely. The final deformed shape shows the "spring-back" of the composite part. And since we did not see any large variation in the temperature distribution from the thermo-chemical analysis we will just assign the temperature amplitude we created to the entire model by modifying the initial predefined field which will be discussed next. For this analysis we will be evaluating three different spar composite thicknesses as shown in Fig. A33 in order to understand the change in spring-back angle for each case.

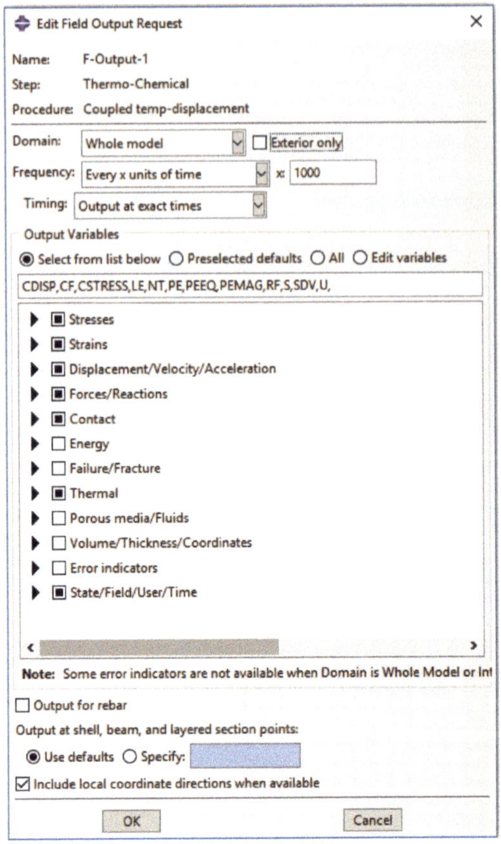

FIGURE A28 Field output request.

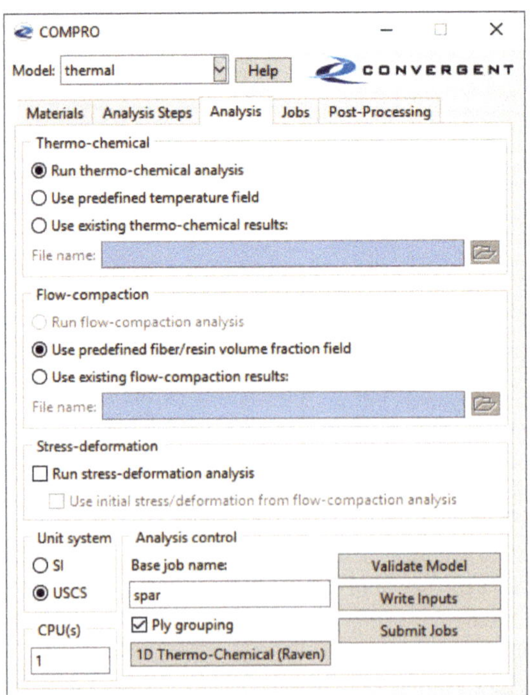

FIGURE A29 Running the model.

FIGURE A30 Model showing the temperature distribution at the end of the cure cycle.

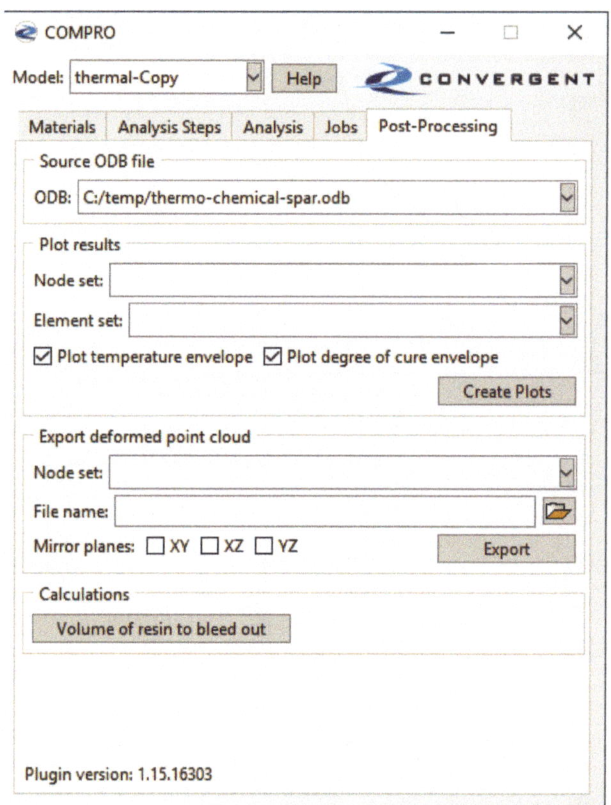

FIGURE A31 Steps to generate the temperature envelop and DOC plots.

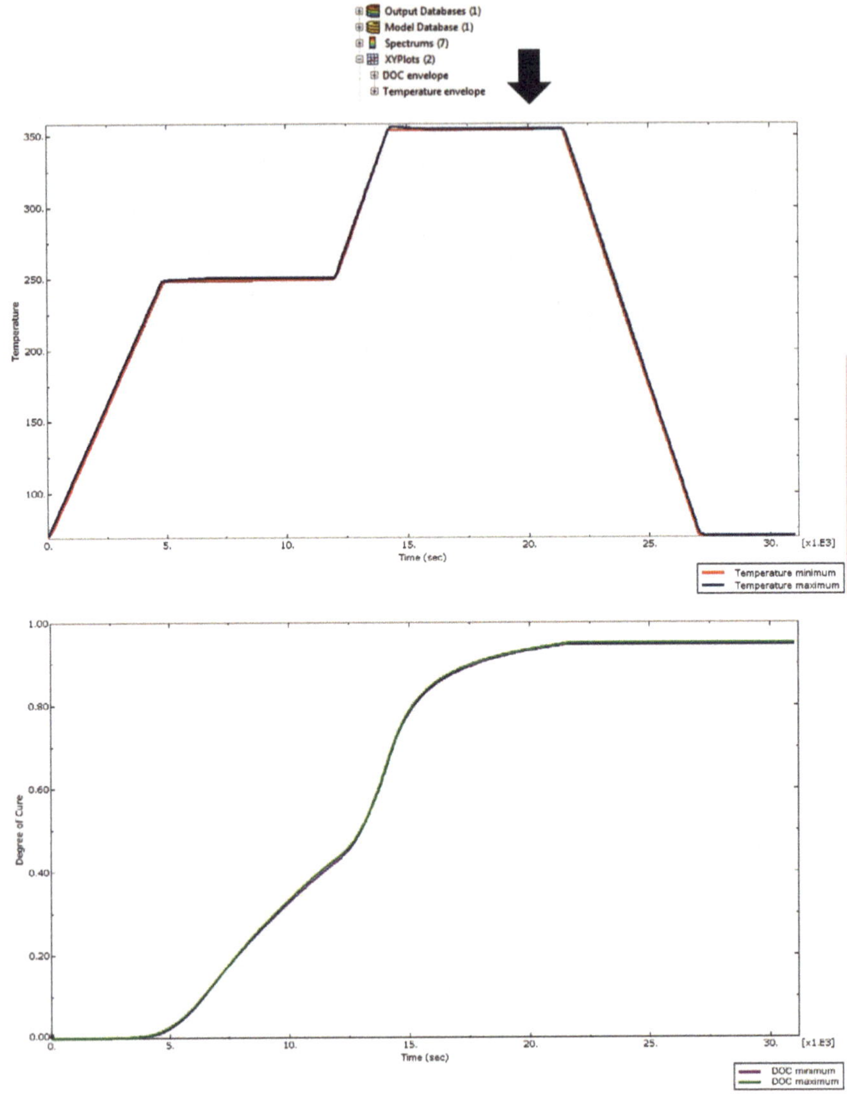

FIGURE A32 Temperature envelop and degree of cure for the model. No large variations are observed in the analysis.

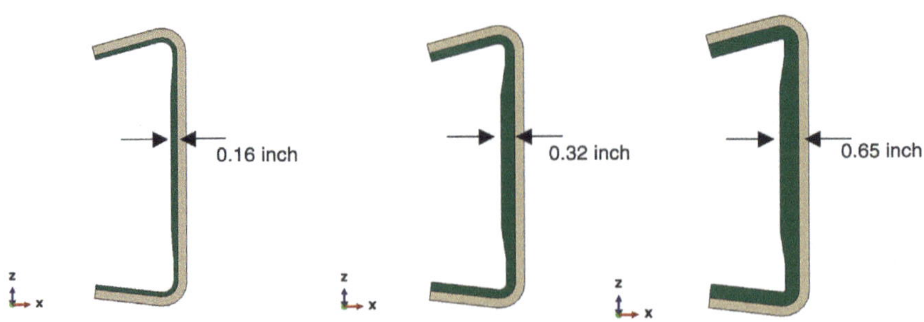

FIGURE A33 Three different spar thicknesses for the stress analysis evaluation.

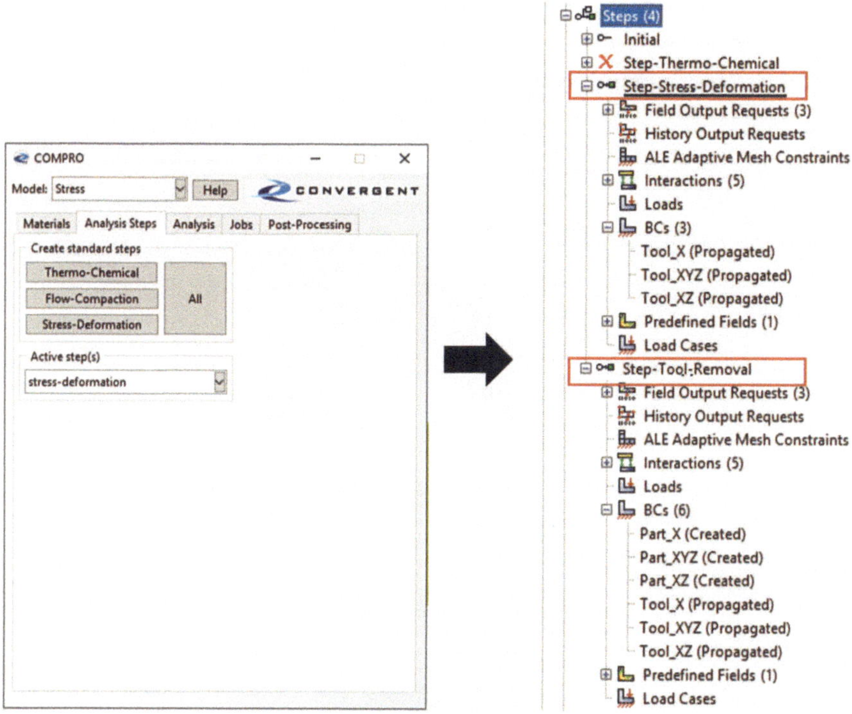

FIGURE A34 Creating a stress-deformation step. Note that once you select this analysis two steps are created one for the actual stress deformation analysis and another to represent the removal of the tool.

Step 26: Create a new stress-deformation step using the plug-in as shown in Fig. A34.
Step 27: In this analysis we will be applying both temperature and pressure simulating a full autoclave cure. Create a surface to apply the pressure to. The surface should be all the exterior surfaces of the model as shown in Fig. A35.
Step 28: Create a pressure amplitude as shown in Fig. A36. The max pressure will be 45 psi.
Step 29: Apply the pressure amplitude to the exterior surfaces of the part as shown in Fig. A37. Note that the magnitude used is equal to 1 and the amplitude selected is the pressure we defined in the previous step.
Step 30: Make sure that the stress deformation step has the correct time period to match that of the cure cycle and pressure amplitude periods used as shown in Fig. A38.
Step 31: In the predefined field under the stress deformation step modify the temperature in order to select the temperature amplitude as shown in Fig. A39.
Step 32: Apply the appropriate boundary conditions in the model. In the stress deformation step we need to apply boundary conditions that will allow the

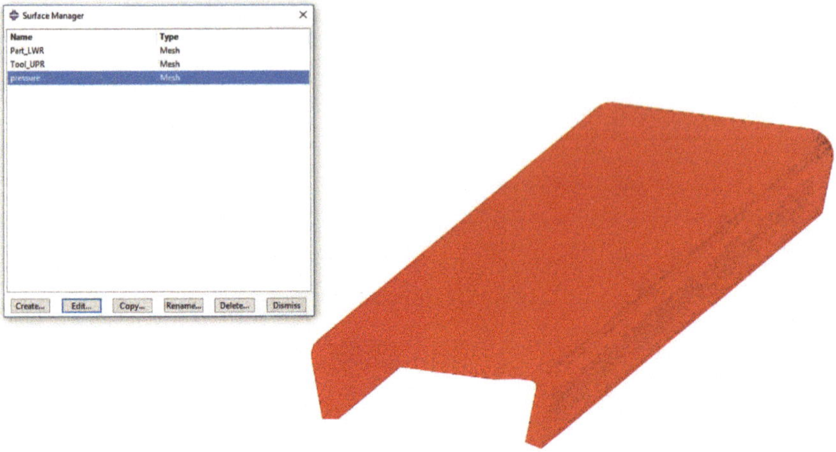

FIGURE A35 Surface consistent of all the exterior surfaces of the model where pressure will be applied.

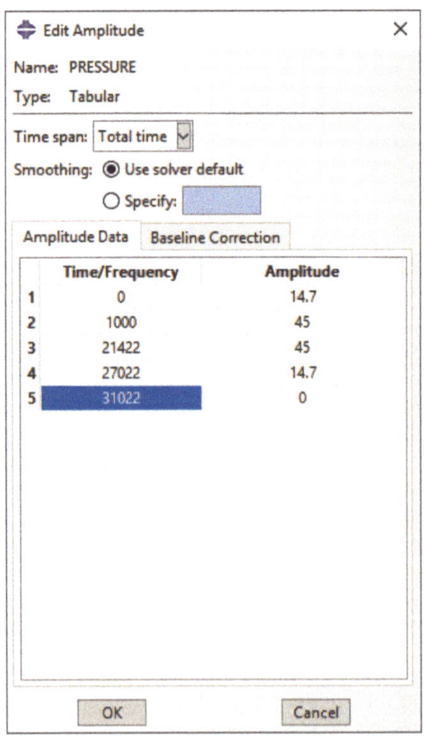

FIGURE A36 Pressure amplitude that will be used during the simulation.

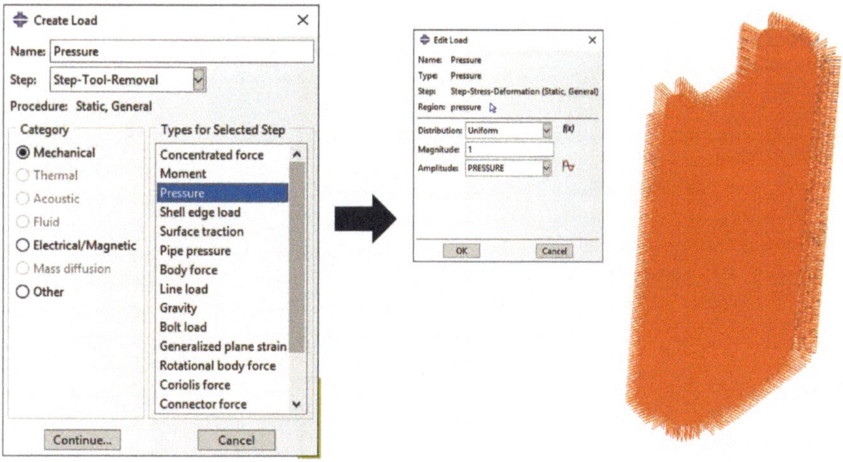

FIGURE A37 Pressure applied to the model.

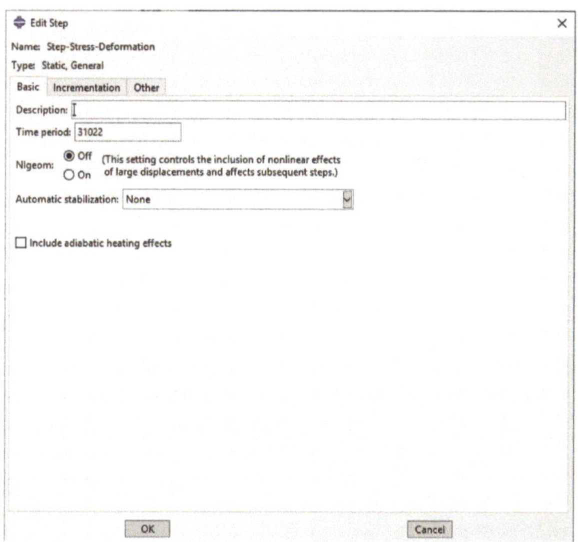

FIGURE A38 Time period used for the stress deformation step.

model to deform without over constraining it. Those boundaries are shown in Fig. A40. All boundary conditions used here are based on the global coordinate system.

Step 33: In the tool separation step make sure that the part_tool interaction created previously is deactivated which ensures that the tool will separate from the part and the free deformation of the part will be evaluated as shown in Fig. A41.

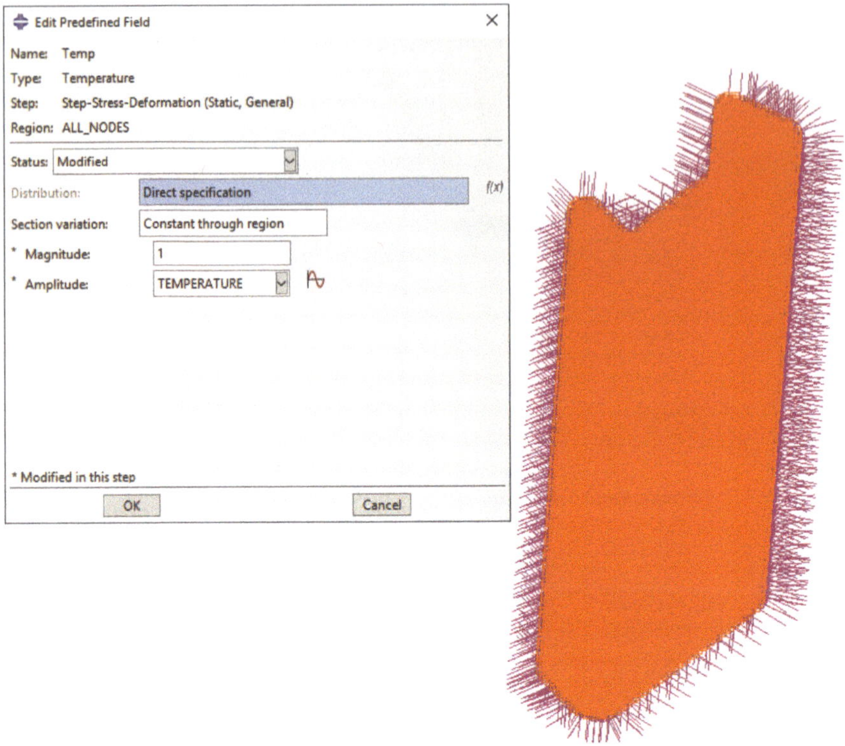

FIGURE A39 Modification of the predefined field to select the temperature amplitude.

Step 34: Now that the part and tool are no longer attached to each other via the interaction property in the tool removal step we will need to apply boundary conditions to the part as to ensure it does not fly away during that step. For that we apply similar boundary as was done previously except that we ensure it is fixed at current position as shown in Fig. A42.
Step 35: Run the simulation. This time we will use the command window to run the job rather than using the plug-in. Create an input file of the model by going to the job module and create job as shown in Fig. A43.
Step 36: Make sure that the input file has the appropriate keywords that are highlighted in Fig. A44.
Step 37: The cca-configuration should be in the same file directory where you will be running the job and include all the materials used in the model as shown in Fig. A45.
Step 38: Run the job by opening a command prompt as shown in Fig. A46. Ensure you are in the correct file directory and write the keywords shown there.

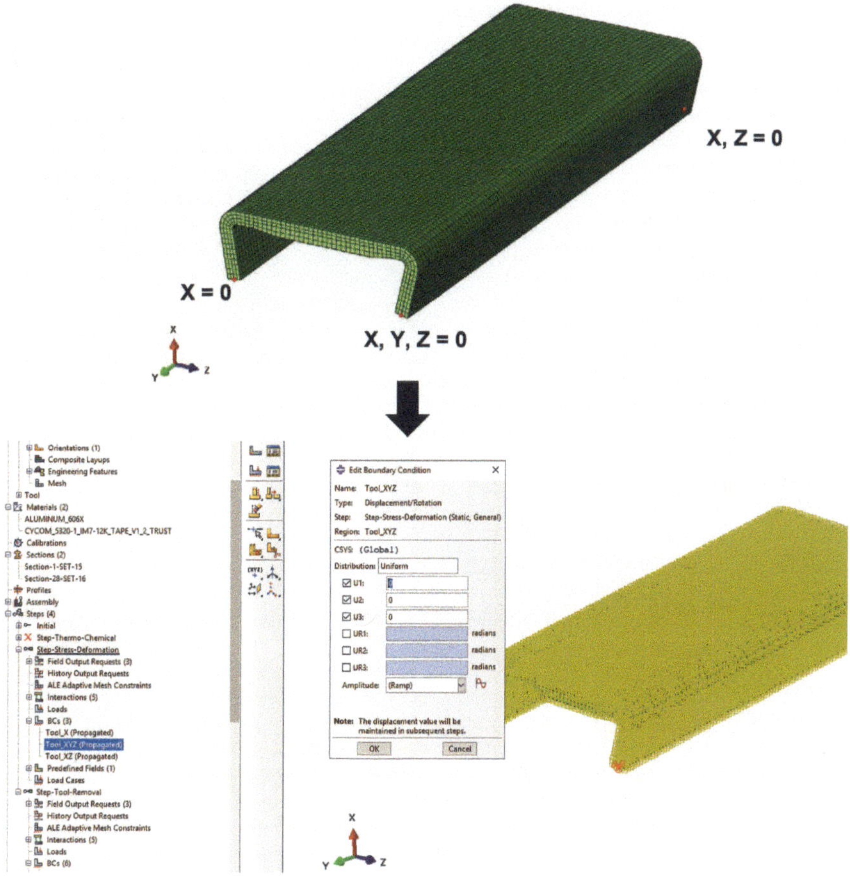

FIGURE A40 Boundary conditions applied to the model.

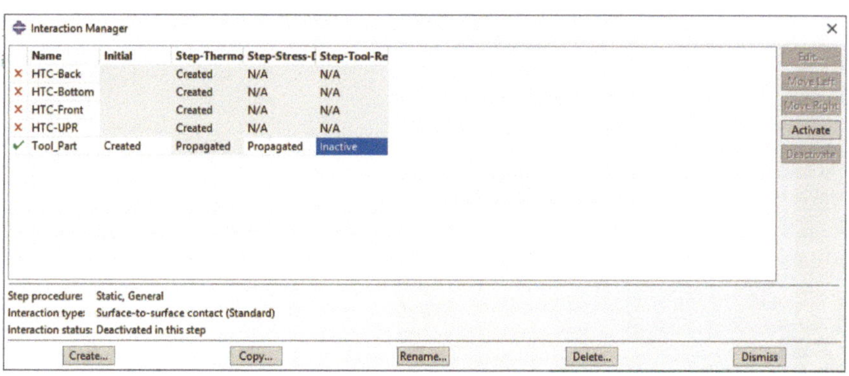

FIGURE A41 Deactivate the part_tool interaction in the tool removal step.

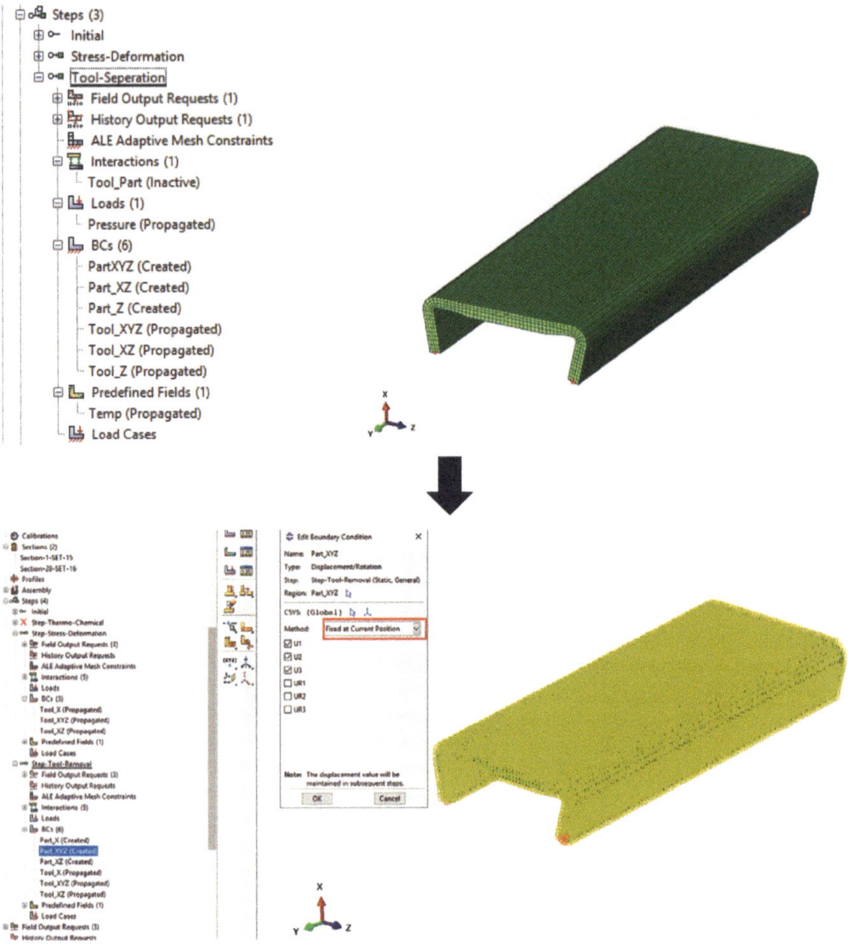

FIGURE A42 Boundary conditions applied during the tool removal step.

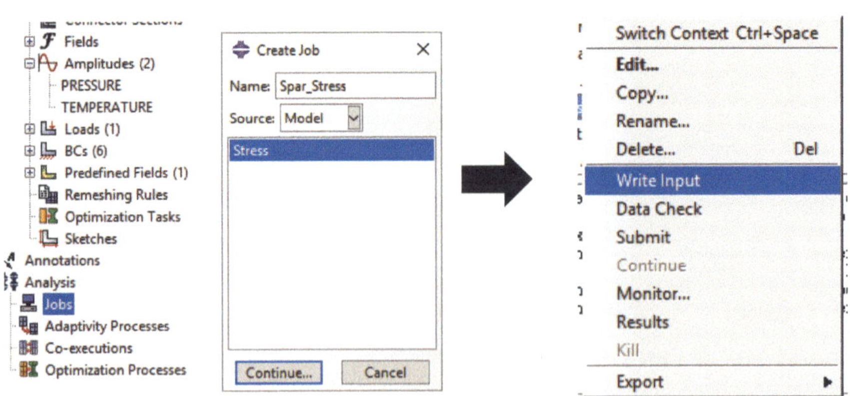

FIGURE A43 Create job and input file based on the generated model.

1.4 Stress-deformation analysis results

Once the analysis has completed we can go to the end of the tool removal step and take a look at the final deformation of the part. Fig. A47 shows an overview of the model at the end of that step.

Since we are interested to see the spring-back values of the model we can isolate the part and interrogate the deformation of each part separately as shown in Fig. A48. As observed, the overall deformation is fairly close between all the models but the spring-back appears to be larger for the thick spar. Fig. A49 shows the comparison between the model thickness and spring-back angle computed in the model. This type of analysis is very useful to generate such design plots to aid the design team in making the correct decisions when it comes to tool design as an example for the appropriate compensation angle to use. It can

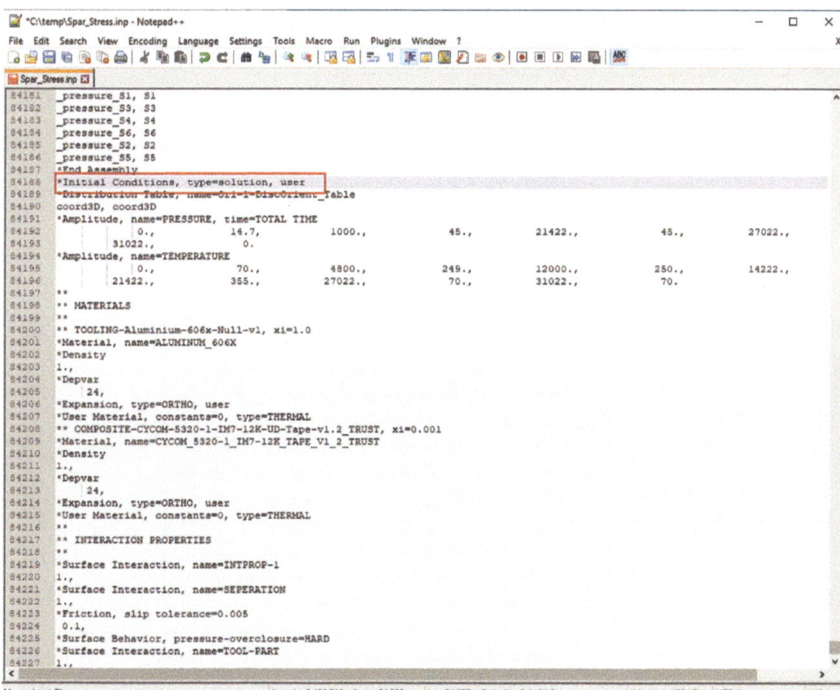

FIGURE A44 Keywords needed to ensure that the Compro plug-in will work appropriately when you run the simulation.

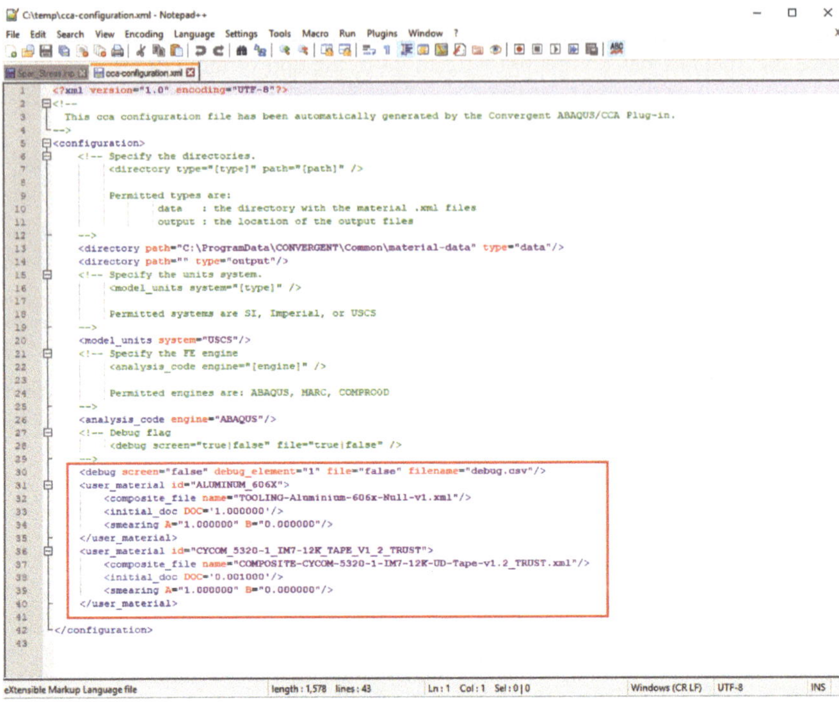

FIGURE A45 Check the cc-configuration file to make sure the materials used are included in the file.

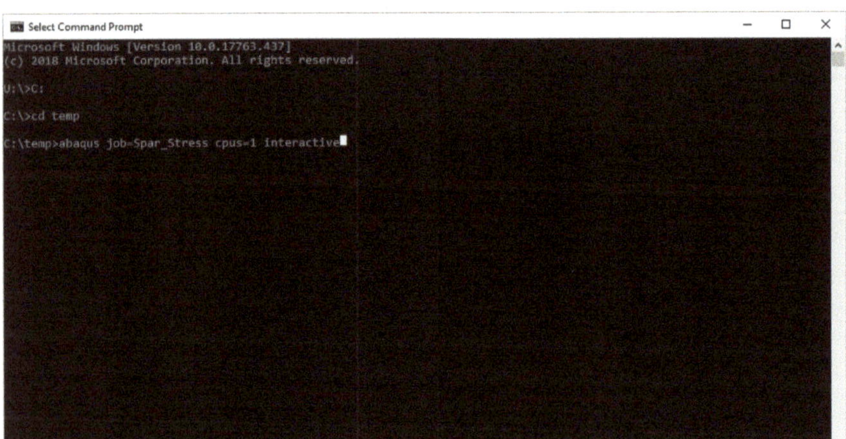

FIGURE A46 Running the job using the command prompt.

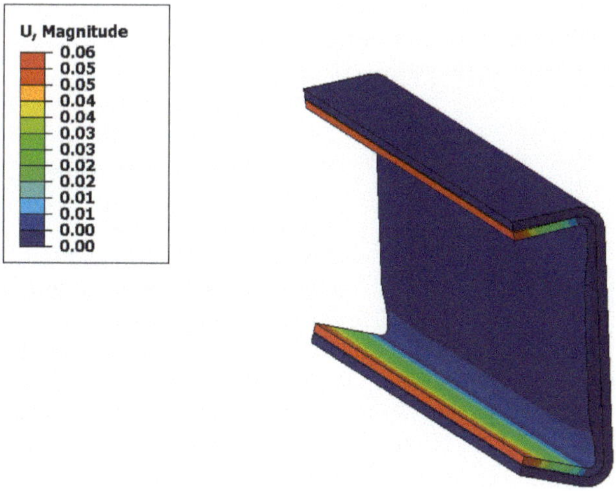

FIGURE A47 Global displacement of the model at the end of the tool removal step.

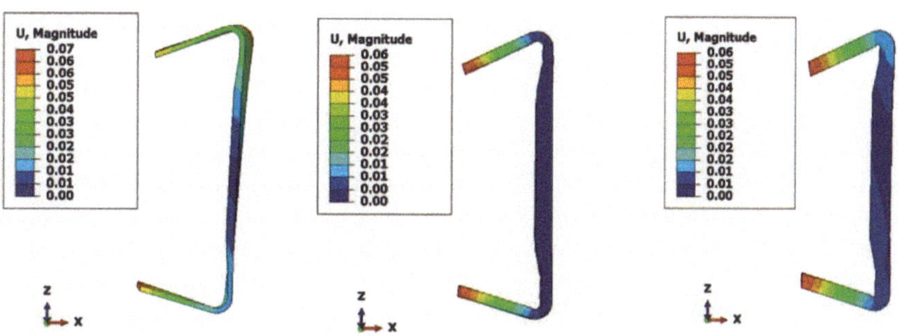

FIGURE A48 Cross section analysis of the different spar designs.

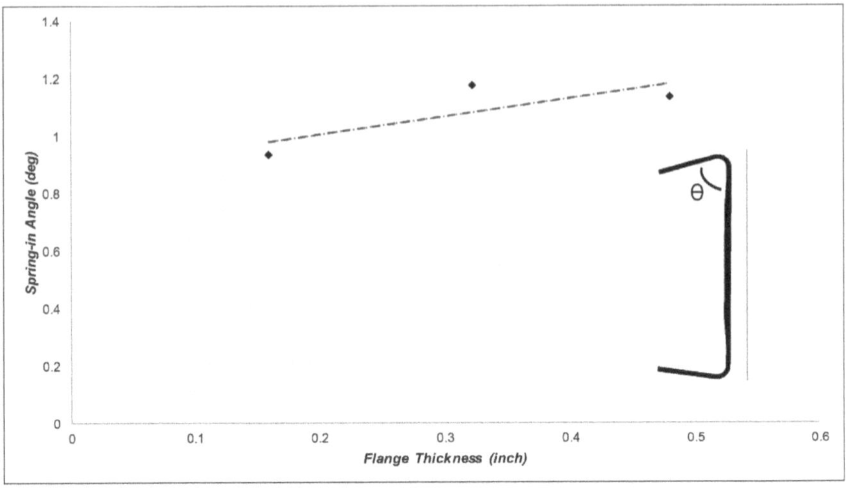

FIGURE A49 Part spring-back for spar like structure as function of flange thickness.

also support the design team in re-evaluating the original design and reducing thickness if the spring-back was deemed high. Note that the deformation shown in Fig. A49 is 20X exaggerated for clarity purposes.

Questions

Q1. Repeat the example discussed in this appendix and use a different tool material and composite material of your choice and compare with the data presented here? What do you observe?

Index

Note: Page numbers followed by "f" indicate figures, "t" indicate tables.

K

L

M

N

O

P

U

V

W

Y

FSC
www.fsc.org
MIX
Papier | Fördert gute Waldnutzung
FSC® C083411

Kolibri 360 GmbH
Ferdinand-Jühlke-Straße 7
99095 Erfurt, Deutschland
produktsicherheit@kolibri360.de